Introduction

Like many people who grew up in a world populated by SR-71 Blackbirds, I've long been fascinated by the aircraft and was disheartened when it was retired. At the time, it seemed that the US Air Force could not possibly have thrown the SR-71 away without having a replacement ready to go; as a result, stories about the hypothetical hypersonic 'Aurora' aircraft seemed to have a fair amount of merit. In the years since (the many, *many* years), though, it has become obvious that no such aircraft as the Aurora exists or ever did exist. So rather than being one more step on an ever-improving capability, the SR-71 was the peak – a peak reached nearly 60 years ago at the time of writing and with no equal in sight. That's the sort of thing that can prove to be a bit of a bummer to an aerospace engineer like myself.

The fact that it looked so unusual, and performed so fantastically, led some to think that its development must have involved the assistance of space aliens or their technology. This belief comes in no small part from the fact that the SR-71 seemed to appear from nowhere and has since had no equal. Without a history, a visible evolution of the design, it is easy to imagine fantastical things. But the fact that the antecedent designs were secret doesn't mean they didn't exist.

History is full of technological innovations that seem to come from nowhere. If you were not involved in the development of, say, the atomic bomb, the transistor, stealth technology or the laser, their advent might have seemed sudden and singular. But they, like the SR-71, had long and convoluted gestations.

In recent years, designs that had been hidden from the public have been declassified and revealed, allowing at least a part of the story of the Blackbird to be told. The tale was told in detail in Paul Suhler's *From RAINBOW to GUSTO*, but until now there has not been a clear visual summary of the many designs that preceded and followed the SR-71. A guide for not only historians and engineers but also aviation enthusiasts and model builders was something I decided that I'd like to see – and that's exactly what you see before you now.

Available illustrations of relevant designs run the gamut from clear, detailed and high resolution, to... well, 'crummy' sometimes doesn't quite do them justice. I strove to create diagrams that were clear, accurate and visually appealing. But in some cases, artistic licence was required. Readers can see for themselves how closely these diagrams follow the originals using the source grading provided in the main data table at the back of this publication

Scott Lowther
aerospaceprojectsreview.com

Contents

Author/artist: Scott Lowther
Publisher: Steve O'Hara
Published by:
Mortons Media Group Ltd,
Media Centre, Morton Way, Horncastle,
Lincolnshire LN9 6JR, Tel. 01507 529529

Typeset by: Burda Druck India Pvt. Ltd.
Printed by: William Gibbons and Sons, Wolverhampton

Acknowledgements: This book would not have been possible without the assistance of Paul Suhler, Tony Landis and Dennis Jenkins. Their assistance and encouragement is gratefully acknowledged.

ISBN: 978-1-911639-72-5

1 CHAPTER Suntan

The 1950s was a time of great advances in aeronautics, with materials science, aerodynamics and propulsion technologies all leaping forward at a pace almost too fast to understand. Only a few years earlier, a world war had begun with fabric-covered biplanes still in service; it had ended with jet fighters and ballistic missiles that flew through space. The sound barrier had recently been broken, and was now being surpassed in multiples. People who knew what they were talking about were talking about sending men to the moon. Aircraft that were just coming off the assembly lines were already obsolete; planning for the future was a mixture of whiz-bang science fiction and blue-sky optimism. So what had once seemed impossible was expected to soon be within reach.

Lockheed's Skunk Works under the management of Clarence 'Kelly' Johnson had designed and produced the U-2 spy plane for the CIA. This was a seemingly simple and straightforward aircraft. Beginning its design life as little more than an extended-span F-104 Starfighter, it proved to be a reconnaissance platform capable of flying so high that neither Soviet aircraft nor missiles could intercept it. And so long as it could not be intercepted, it could be denied… even if the Soviets could clearly see it far above their heads.

But just as aeronautics advanced in the United States, so too did it advance in the Soviet Union. Some in the CIA and USAF could see that the Soviets would soon develop systems able to bring down a U-2, at which point it would become problematic. Fortunately it was understood that the U-2 could be replaced by something capable not only of cruising at great altitude, but also at a far higher speed. What technologies would permit this took some time to determine and a variety of different options were explored.

Throughout the 1950s, a great deal of time, effort and taxpayer dollars were expended in the eventually futile effort to produce atomic powered aircraft. That an aircraft could be powered by atomic energy had been assumed since the Second World War, but actually achieving that capability proved far more difficult and dangerous than originally thought. Many aircraft and engine designs were put forward, with craft such as the Convair X-6 (a highly modified B-36 with four reactor-powered turbojets) and the Convair NX2 gaining a fair amount of press. Additionally, popular magazine articles abounded with chatter about A-Planes, and from the vantage point of 60-70 years later it certainly looks like many thought aircraft with atomic engines would soon be zooming through the stratosphere or above. This, of course, was not to be.

While atomic power for aircraft fizzled under the glare of the public spotlight, another technological approach was carried out relatively quietly. Aircraft to this point had been powered by the burning of hydrocarbons, typically gasoline, kerosene or something similar. These fuels were well known; liquid at room temperature, they could be easily handled in something as simple as a bucket. Many were not easily ignited and were considered quite safe. Chemists, though, knew that something potentially better was possible.

Experiments had begun during the war on the use of liquid hydrogen as a rocket propellant. Burned with liquid oxygen, the resulting exhaust would be water vapour (and if burned fuel-rich, free hydrogen which would then combust with the surrounding air). Compared to hydrocarbons that produced a substantial fraction of carbon dioxide (CO2), the far lower molecular weight of water (H2O) with the not inconsiderable bonus of more energy per unit mass of propellant when compared to hydrocarbons meant theoretically much better performance.

Shortly after the war, North American Aviation, Martin and Aerojet all produced designs for single-stage hydrogen-fuelled rockets capable of putting meaningful payloads into orbit. Contemporary designs of rockets using gasoline or other hydrocarbons for fuel were much heavier (and thus costly) and required more stages (and thus even more costly).

While early work with liquid hydrogen rocket fuel demonstrated its performance potential, it also had certain rather dire problems: firstly, it was an extreme cryogen. To remain liquid at low pressure, hydrogen must be cooled to below -423°F (-253°C). Further, the density of liquid hydrogen is fabulously low, some seven percent that of water, with the result that a rocket equipped with liquid hydrogen would have to be far bigger and more voluminous than its equivalent hydrocarbon-fuelled stablemate. But even with greatly increased tank sizes, the fully fuelled weight of the hydrogen rocket would be well below that of the hydrocarbon rocket, meaning fewer and/or smaller rocket engines would be needed.

Using hydrogen as an aircraft fuel certainly posed challenges – it wouldn't be as simple as switching out the fuel tanks – but it also created opportunities.

Hydrogen ignites very rapidly with air and disperses quickly due to flash evaporation, meaning that complete combustion can occur in a shorter distance. This becomes especially apparent at high altitudes with thinner air – therefore turbojets and ramjets designed for hydrogen fuel could be relatively compact compared to their hydrocarbon counterparts. This lowered engine weights and airframe mass while improving combustion efficiency.

And hydrogen will combust just fine from sea level to the stratosphere, so many in the aeronautical industry saw it as a very promising propellant for high-speed high-altitude aircraft. But before aircraft could be designed to make use of this new wonder propellant, engines needed to be designed – at least on paper – that would properly burn it and turn it into thrust. One of the most important of the early concepts was the 'Rex-1' produced by Randy Rae of Summers Gyroscope.

Rae presented the Rex-1 to the Wright Air Development Center in 1954, where it was well received. The engine was meant to power a subsonic high altitude airplane, turning a propeller to generate thrust in the normal way… but the way it turned the propeller was new. The engine did not rely on atmospheric oxygen at all, instead burning hydrogen and oxygen in a more or less conventional rocket engine. The rocket burned fuel-rich to keep the exhaust temperature low enough so that a turbine located behind the exhaust would not be damaged. Behind that turbine, the exhaust gases had additional oxygen added, to continue the combustion process, and passed through another turbine. And once more oxygen was added and once more the gas passed through a turbine. The exhaust then passed through a heat exchanger to heat incoming hydrogen fuel before injection into the rocket engine.

By turning a three-stage turbine, the Rex-1 converted the high exhaust velocity of a rocket engine into torque for a propeller. The 'exhaust velocity' created by the propeller pushing air was very low compared to that of a straight rocket engine but the mass flow was vastly higher. The end result was that the effective thrust of the rocket was multiplied.

Analysis showed that the Rex-1 engine would indeed provide thrust for an aircraft and at virtually any altitude. Contemporary jet engines would suffer combustion instability at around 14km altitude, but the Rex-1 would never have that difficulty. The liquid oxygen ensured that the engine would continue to operate and spin the propeller all the way to the vacuum of space if need be. But the fact that the Rex-1 required a tank of liquid oxygen, rather than using atmospheric oxygen, meant that it had its own limitations. As efficient as it was compared to a rocket engine, it fell short compared to a conventional turboprop engine. It was also limited to a strictly subsonic aircraft.

However, by early 1955 the Rex system had been greatly revised, with three new engines envisioned. The Rex I was now a jet engine; it still burned liquid oxygen with the liquid hydrogen but the propeller had vanished. The combustor drove a jet engine-like compressor and simply exhausted the hydrogen-rich gas out the rear of the engine. The Rex II used that hydrogen rich gas in an afterburner, obtaining useful thrust from the waste gas. The Rex III was an all-new engine; liquid oxygen was removed entirely, the engine was a full airbreather using hydrogen fuel. The liquid hydrogen was pumped at high pressure through a heat exchanger at the rear of the engine, where the liquid was heated to high pressure gas by hydrogen/air combustion. The hydrogen gas was then used to drive the turbines in a three-stage process, as was done with the original Rex engine. And at the end the heated hydrogen gas was burned with air in combustors to power the heat exchangers in the first place.

At this point, the Rex engines were still thought of as engines for subsonic aircraft. But what many in the Air Force and the NACA (National Advisory Committee for Aeronautics) really wanted was a turbojet that could burn hydrogen and do so not only at high altitude but also high speed.

Studies at the NACA Lewis Flight Propulsion Laboratory in late 1954 showed that a hydrogen-fuelled turbojet should not suffer combustion instability until more than 30km altitude, twice what contemporary conventional jet engines could handle. By mid-1955 conceptual design work at Lewis had produced simple concepts for large hydrogen-fuelled, jet-propelled high-altitude reconnaissance, bombardment and fighter aircraft using cylindrical insulated tanks. With a configuration approximating that of a B-52 bomber, a subsonic configuration was produced for both bombardment and reconnaissance roles, with the fuselage largely composed of hydrogen tanks and smaller diameter, long, thin tanks filling up much of the space in the wings.

For supersonic roles, the fuselages were long and spindle-shaped; the wings were much like those of the F-104 – thin, stubby, unswept and devoid of propellant tanks. Simple illustrations were provided of the subsonic aircraft and the supersonic bomber and fighter configurations; the supersonic reconnaissance aircraft was not illustrated but is here reconstructed from data provided and a claimed similarity to the bomber configuration.

The supersonic aircraft were intentionally designed to be geometrically simple. Complex organic curves, wings smoothly blended into the fuselage, non-circular cross-sections… these would all greatly exaggerate the already substantial difficulties of designing for liquid hydrogen. So the fuselages of the supersonic aircraft were simple bodies of rotation, just the thing for a long vacuum-walled 'Thermos bottle' cryogenic liquid fuel tank. Surface area and structural mass were minimized and the structure itself was reasonably strong. The tanks would be made from a thin metal such as stainless steel, and as with the Atlas ICBM internal pressurization would be used to maintain rigidity of the tanks and to turn the tanks into important structural elements. Around the exterior of the stainless steel 'balloon' there would be several inches of a plastic foam insulation, which would be covered in an aluminum foil.

The supersonic reconnaissance aircraft was designed to have a maximum weight of 75,000lb and to fly at an altitude of 80,000ft at Mach 2.5. It was expected that it would climb subsonically to 40,000ft, accelerate to Mach 2.5, then continue climbing to 70,000ft. At that point it would cruise at a constant speed while slowly climbing to 80,000ft as it lost fuel weight. Once at that speed and altitude it would stay there for much of the rest of the mission.

The concept was quite simplistic, and various design elements were adjusted to see what effect they would have on performance. As intriguing as the ideas were, these and related NACA-Lewis hydrogen-fuelled designs seemed to fade away. To those not in the know, this doubtless would have looked like just another example of an interesting 1950s aviation idea that was passed over.

At about the same time that NACA-Lewis was designing its hydrogen-burning aircraft, Lockheed entered the story. In October 1955, Garrett Corporation, which had purchased Summers Gyroscope's interest in the Rex engines and technologies, contracted with Lockheed to design an aircraft to go with its new 48in diameter Rex-III engine. The USAF had become interested in a hydrogen-fuelled reconnaissance aircraft using the Rex engine, but did not feel that Garrett should be responsible for both the engine and the airframe. Kelly Johnson of the Lockheed Skunk Works led the work that produced the CL-325-1 design in January 1956.

CL-325-1 looked not unlike the NACA-Lewis design, just with parts shuffled around and resized somewhat. It also very clearly took much of its appearance from the contemporary Lockheed F-104, produced by many of the same Skunk Works designers. It used a spindle shaped fuselage composed almost entirely of a liquid hydrogen fuel tank; straight F-104-like wings with noticeable anhedral; two large wingtip-mounted engine nacelles; and a large F-104-like T-tail. The single pilot cockpit was well forward and looked quite small compared to the aircraft as a whole. Avionics and reconnaissance equipment were all located in the nose, along with landing gear. Main landing gear folded up into the thin biconvex wings. Construction was to be largely of aluminium.

A second design was shown at the same time. The CL-325-2 was designed along the same lines as the CL-325-1 but with jettisonable external tanks. This

NACA Hydrogen Fuelled Photo Recon

SCALE 1/300

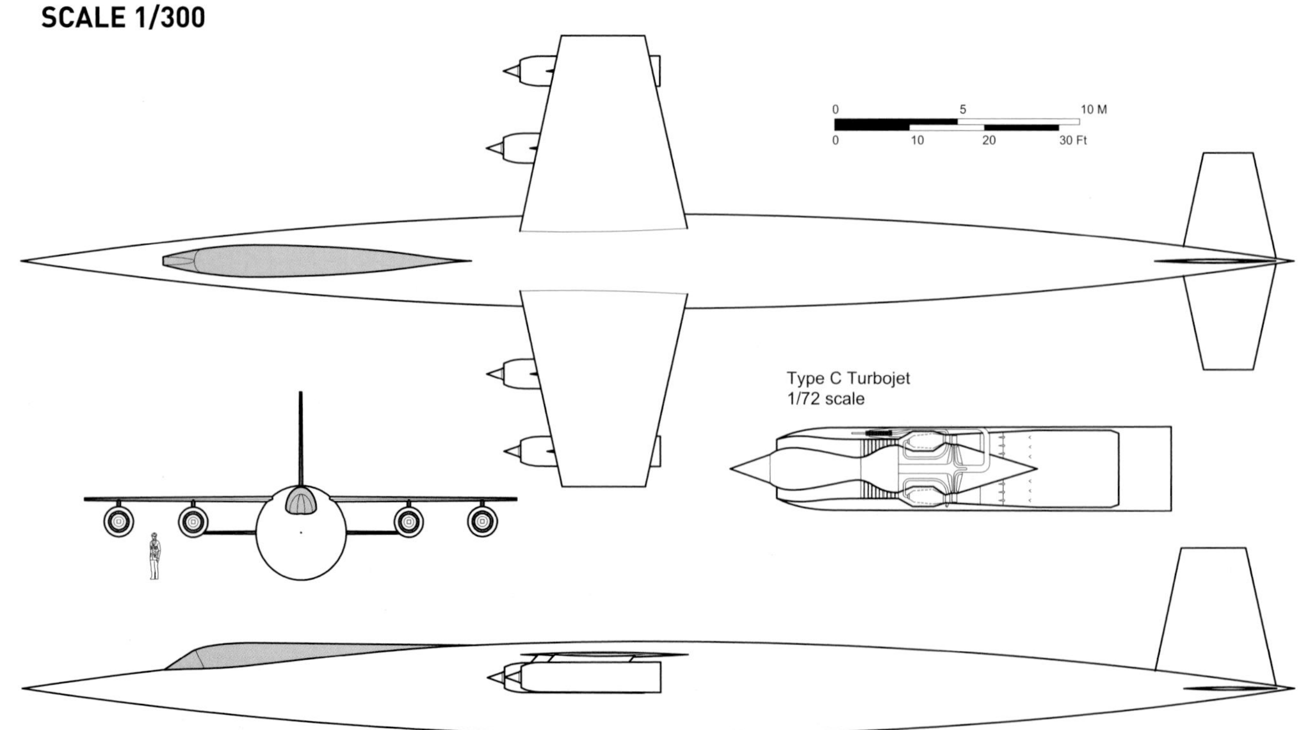

Lockheed CL-325-1

SCALE 1/275

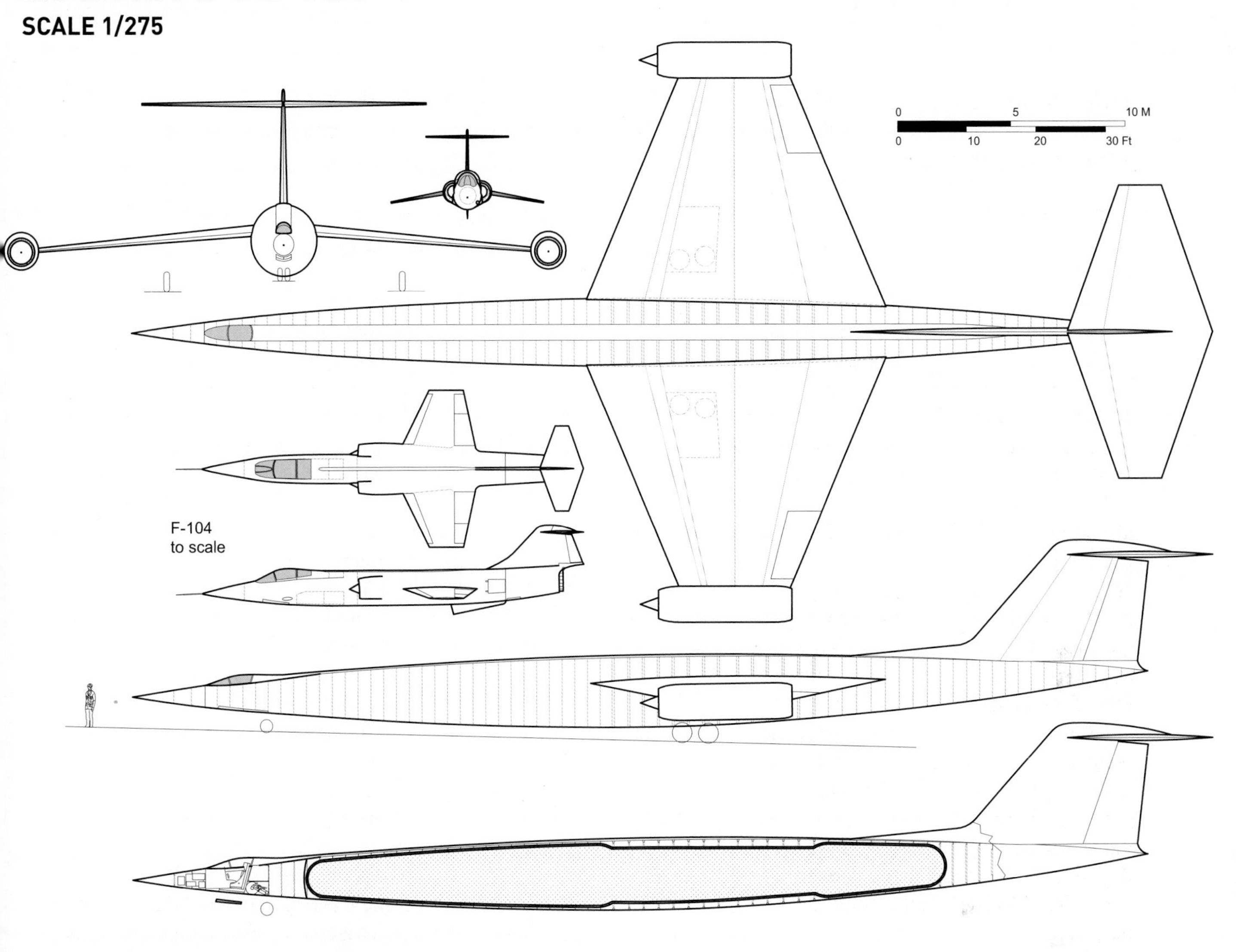

Lockheed CL-325-2

SCALE 1/275

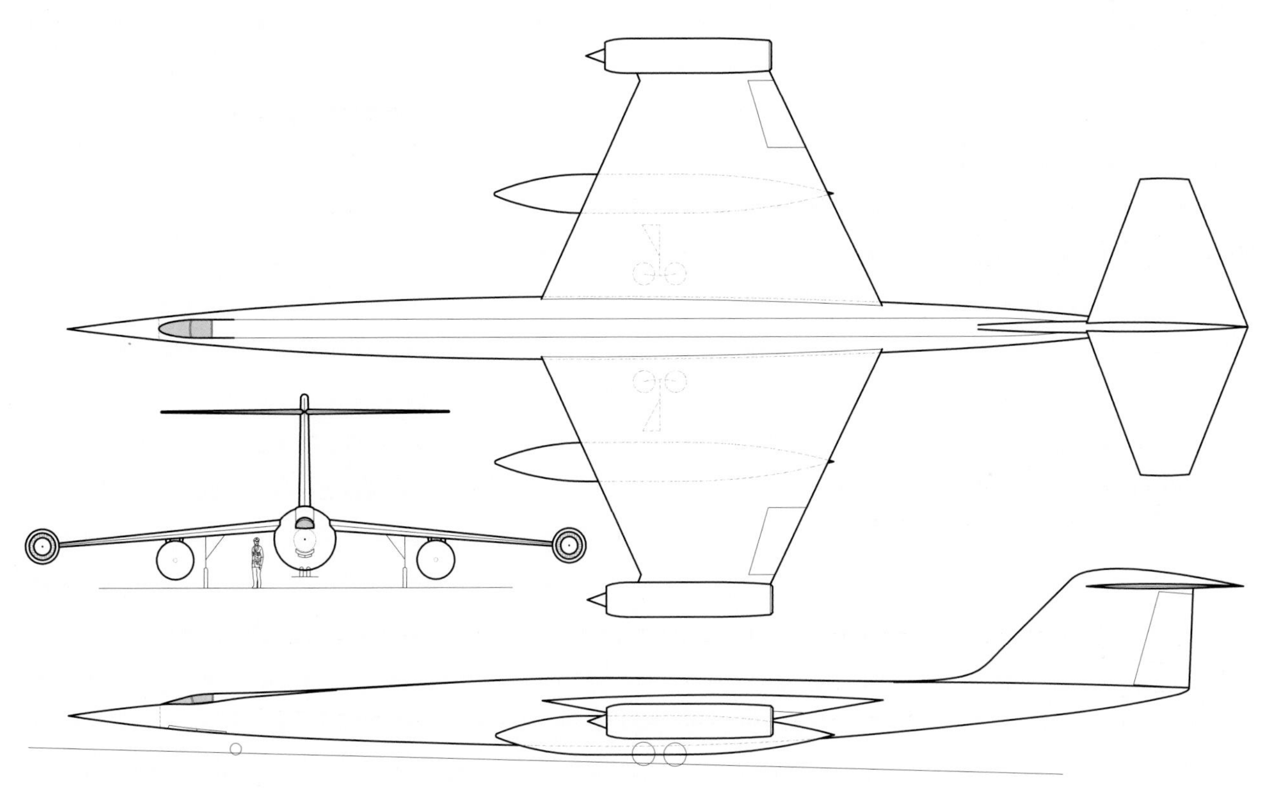

resulted in an aircraft of somewhat smaller dimensions but greater gross weight (note that the available data on weights adds up to a greater sum than the total given), and of course the greater expense associated with each flight requiring two brand-new gigantic liquid hydrogen tanks.

The Garrett team turned in their first report, including the Lockheed report as an appendix, in February of 1956 to what they thought was a positive reception. However, the Rex III and the CL-325 were larger and more complex designs than the Air Force had expected. The Air Force decided that Garrett did not have the experience or infrastructure needed to completely develop a full scale engine. The Rex III was a much larger engine than any Garrett had worked on by that time and they did not have the physical infrastructure in place to work with large amounts of liquid hydrogen. So, much to the Garrett team's surprise, their efforts were passed up. Existing contracts to study the engine continued, with final reports issued in 1958, but their hopes of being tapped to fully develop the engine were dashed.

While Garrett was essentially out of the picture, other companies were not. In particular, Kelly Johnson had gone directly to the Air Force with the concept in January of 1956. Some in the Air Force were still stung by their passing on the U-2, which the CIA had snapped up. They were ready to jump on board the next Lockheed concept, this time for a higher flying and much faster recon aircraft. Johnson's fast and reliable work on the U-2 and other aircraft led the Air Force to believe that he could quickly accomplish the goal of a fast, high altitude hydrogen-fuelled reconnaissance aircraft. And so while Garrett were still vainly trying to sell their engine concept to the Air Force, Lockheed began work on their airframe. The Garrett Rex III engines were replaced with all new engines.

In January of 1956 Pratt & Whitney and General Electric were asked to turn in engine proposals, and by the end of February – about the same time Garrett was turning in their report on the Rex III – Pratt & Whitney's proposal was selected. The Pratt & Whitney 304 expander cycle turbojet was quite similar to the Rex III in concept, but it had the advantage of coming from a known manufacturer of large engines. Contracts with both Pratt & Whitney and Lockheed were signed in May of 1956. The efforts, especially Lockheed's, were given high security classifications. It took the better part of two decades before any information on the new aircraft became public.

As the design was evolved and refined, the CL-325-1 quickly transformed into the CL-400. Around this time the project was given the code name 'Suntan,' appropriate given that the sun is fuelled by hydrogen. Lockheed agreed to build two prototypes within 18 months of go-ahead. The new design was quite similar to the old, but with some substantial changes. The most obvious change was the addition of a ventral fin, the additional stability it provided allowing the anhedral of the CL-325's main wing to be removed. For landing the ventral fin would fold to the side. The fuselage remained a long body of rotation, but ended up more cylindrical than the spindle-shaped fuselage of the CL-325. Where the CL-325 had a single long hydrogen tank, the CL-400 had three… a long one in the forward fuselage (17,800 gallons), a long one in the rear fuselage (14,240 gallons) and a smaller sump tank (4000 gallons) in the middle. This would be heavier and less volumetrically efficient, but it would allow greater control over centre of gravity issues. To aid fuel transfer between tanks, the two main tanks were kept at 19 psig and the sump tank at 17 psig. For fuel transfer to the engines, pumps would push 385 gallons per minute at 50 psig through vacuum jacketed lines through the wings.

The CL-400 had landing gear at the nose and main wheels in the fuselage near the wing spars, creating a bicycle arrangement. Fairings on the undersides of the wingtip engine nacelles covered small outrigger gear. As with the CL-325, construction would have been largely of aluminium. Artwork from the time often depicts the pilot's canopy being one sizable transparency, as it had been for the CL-325, and somewhat similar to the canopy of the F-104. But more detailed art, and examination of the few poor-quality photos that have emerged of the full scale mockup of the CL-400 forward fuselage, show that the canopy was broken up into a number of smaller transparencies. The pilots had, apart from the numerous window frames, a quite good view; the reconnaissance systems officer had a more restricted view through side windows. Small domes on both the top and bottom of the forward fuselage between the pilot and the RSO provided visibility for navigation optics. A pair of sizable cameras were contained in a bay directly aft of the RSO. A raised spine ran the length of the vehicle providing volume for cables and the like to reach from the cockpit to the tail without having to penetrate the full-diameter fuel tanks.

The CL-400, as advanced and well designed as it was, was troubled from the start. The low density and cryogenic nature of liquid hydrogen limited the design options for the aircraft, with the result that range was never what the Air Force wanted. Kelly Johnson became fully aware of this problem during the first six months of the effort and was unable to truly solve the problem. This would have meant that the CL-400 would have to either be refuelled in flight – a difficult challenge with liquid hydrogen, but an option that was studied – or based at airfields quite close to Soviet overflight destinations. This would have meant a substantial liquid hydrogen infrastructure would have had to be built not just within the United States, but in foreign countries. Either substantial local hydrogen generation and liquefaction facilities would have to be built (expensive, dangerous, likely controversial for the locals and readily damaged via either accident or sabotage), or a worldwide liquid hydrogen transportation system would need to be devised. Neither option was appealing.

Problems with range led to interest both at Lockheed and the Air Force waning by the middle of 1957, but in true bureaucratic fashion the doomed programme hung on until officially cancelled in February, 1959. Suntan was not, however, a wasted effort. Lockheed demonstrated the ability to make great progress with advanced aircraft development in short time spans, an ability that was important in the development of the Archangel series and ultimately the A-12. For Pratt & Whitney, the 304 turbojet was never flown… but the lessons learned about how to handle liquid hydrogen – on the ground, in a vehicle, in an engine – paved the way for the development of the RL-10 rocket engine. This engine was flown initially on the Atlas-Centaur and Saturn I launch vehicles and continues to fly, in modified and improved form, on the Atlas V and Delta IV rockets today.

Since the CL-400 proved incapable of attaining the sort of range the Air Force wanted, Lockheed produced a number of advanced concepts building on the technology. The information available on these is fairly lean, coming from a single paper written some years after the fact. These designs seem to have been created around 1957.

The earliest known of the 'improved' CL-400s is the CL-400-10 which was to be geometrically identical to the CL-400. The difference was that it was to have integral tanks, increasing not only structural weight, but also total fuel weight. The resulting range increase was noticeable. The integral tanks must have been considered particularly challenging for this not to have been more highly thought of as a solution to the range problem.

The CL-400-11 was an attempt to extend the range of the aircraft by increasing the size of it. The fuselage was stretched more than forty feet; the dry weight of the

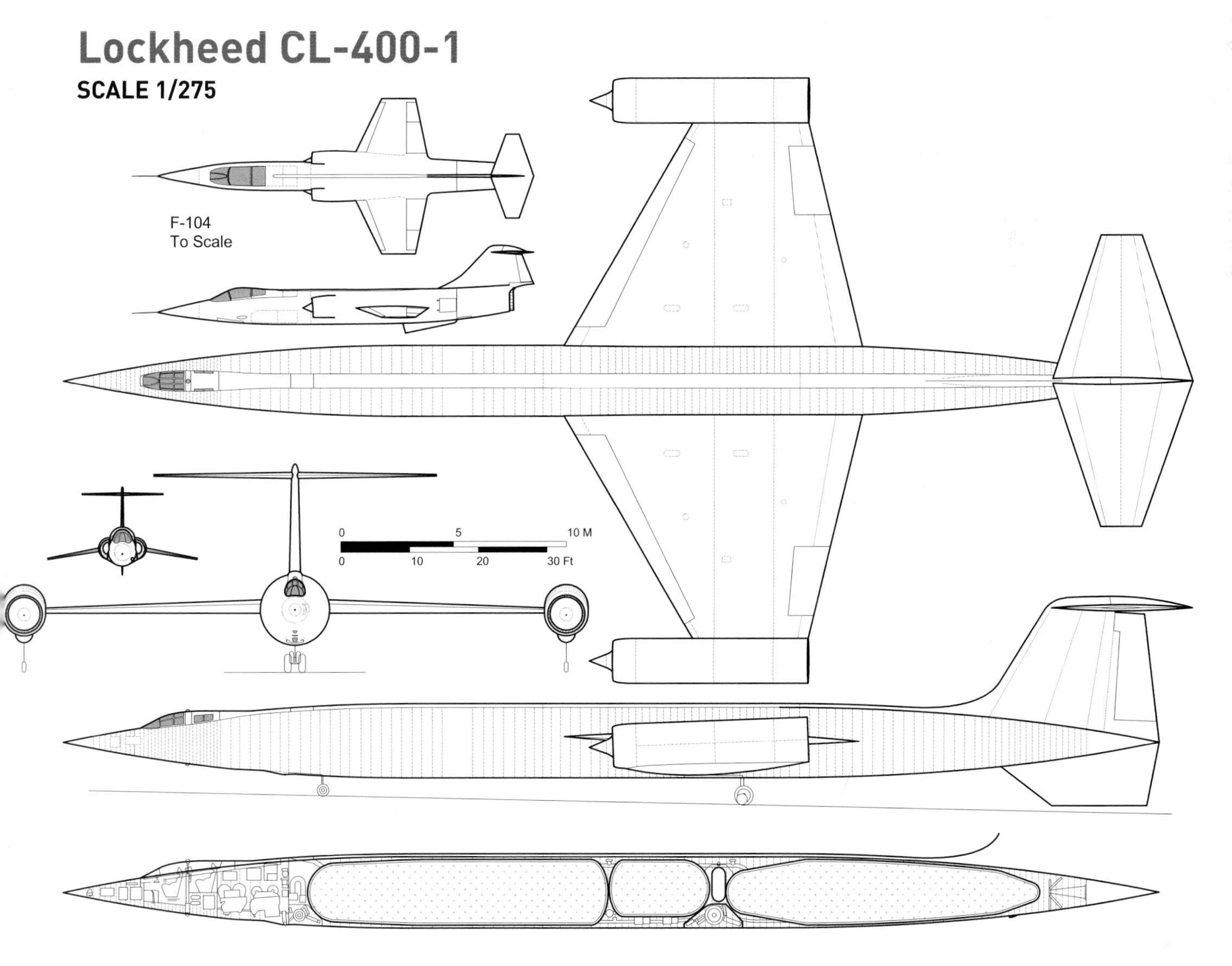

vehicle went up by more than a third. But the hydrogen fuel weight more than doubled. As a result, range nearly doubled. The same engines were used, but they were moved inboard to inside of mid-span. For some reason the aircraft had three separate spines… one on top and one along each side. They are unexplained, but probably indicate that the aircraft truly filled the fuselage with fuel and required external runs for not only cables but perhaps structure as well.

The CL-400-12 was basically a scaled-up CL-400-12, now to 272ft in length. More than a hundred feet longer than the basic CL-400, it required four of the P&W 304-2 engines. Performance in terms of speed and altitude was essentially the same as for the CL-400 (with altitude performance actually being slightly less), but range was improved by more than 2500 nautical miles

The CL-400-13 was a complete redesign for a massive increase in performance. The design was

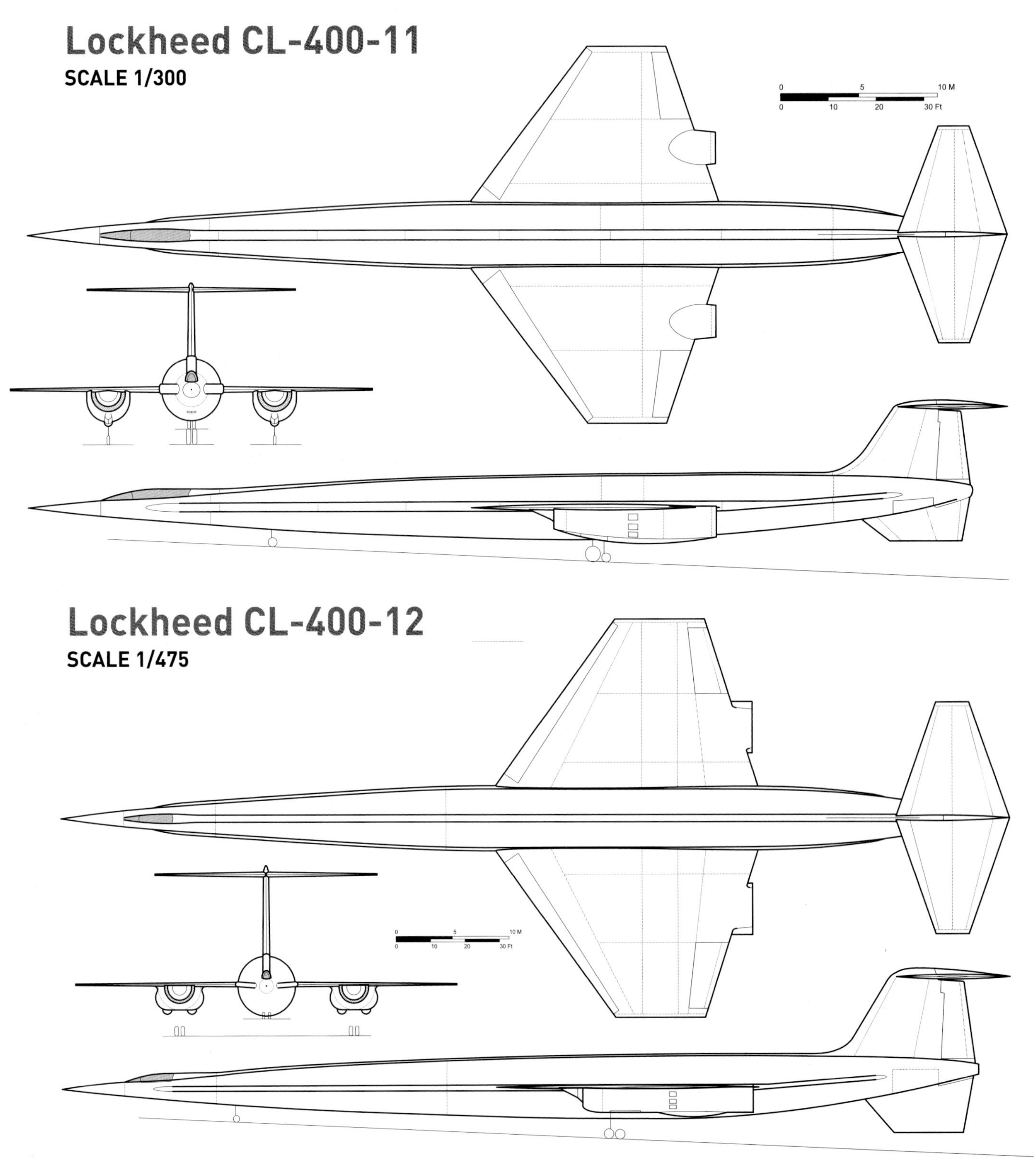

gigantic at nearly 300ft in length; the fuselage is conical and the wings delta, moved to the rear. Small canards provide lift and stabilization. Two new engines (Pratt & Whitney STR-12s scaled up to 125%) were mounted in the lower portion of the rear fuselage. The resulting design looks less like a reconnaissance aircraft and more like a single stage to orbit aerospaceplane... which it may well have been related to given that AeroSpacePlane studies were ongoing at the time. Swap out the expander cycle turbojets and give it a bank of scramjets and it would have comfortably fit in with the X-30 National Aero Space Plane studies of the 1980s. As it was, the design was meant to cruise at Mach 4 for an astounding 9000 nautical miles. The down side was an aircraft of monumental and quite possibly wholly impractical proportions.

The CL-400-14 was a return to the CL-400-11/-12 configuration, but with canards. It remained a very large vehicle at 290ft and used four P&W STR-12 engines scaled to 85% for thrust. Like the -13 it cruised at Mach 4, but range was slightly reduced.

The last of the designs was the CL-400-15JP, for Jet Propellant. This was a design thrown in for comparative purposes, with the troublesome hydrogen fuel replaced with conventional hydrocarbon fuel fuelling a pair of conventional P&W J58 engines. The design seems rather simplistic and crude, but it showed that conventional fuels could get the job done for a whole lot less trouble... and with a more reasonably sized vehicle.

Weapon System 118P

Lockheed was not alone in studying liquid hydrogen fuel for a high altitude, high speed reconnaissance aircraft. Apparently Boeing had a four engined design in direct competition with the CL-400; sadly the only other information currently known about it is that it was to be 200ft long with a delta wing of 200ft span.

But there was another apparently separate programme that seemed to compete with Suntan: Weapon System 118P High Altitude Reconnaissance Program. Begun in September 1955, WS 118P was a multi-phase Air Force (Wright Air Development Center) project to design a series of piloted reconnaissance aircraft. At this time the Lockheed U-2 was so classified that many in the Air Force who would have liked to have had the U-2's capabilities were unaware that the programme existed; so parallel efforts, such as WS 118P, were initiated.

Phase I of the WS 118P is believed to be the Martin B-57D and the Phase II design was the Bell X-16. Both of these were subsonic twin-turbojet designs configured to cruise at high altitude. But the major portion of the HARP programme was aimed at Phase II ½ and Phase III. Phase II ½ was aimed at an aircraft that would cruise above 75,000ft with a range of at least 2400 nautical miles, operational in 1958. Phase III bumped cruise

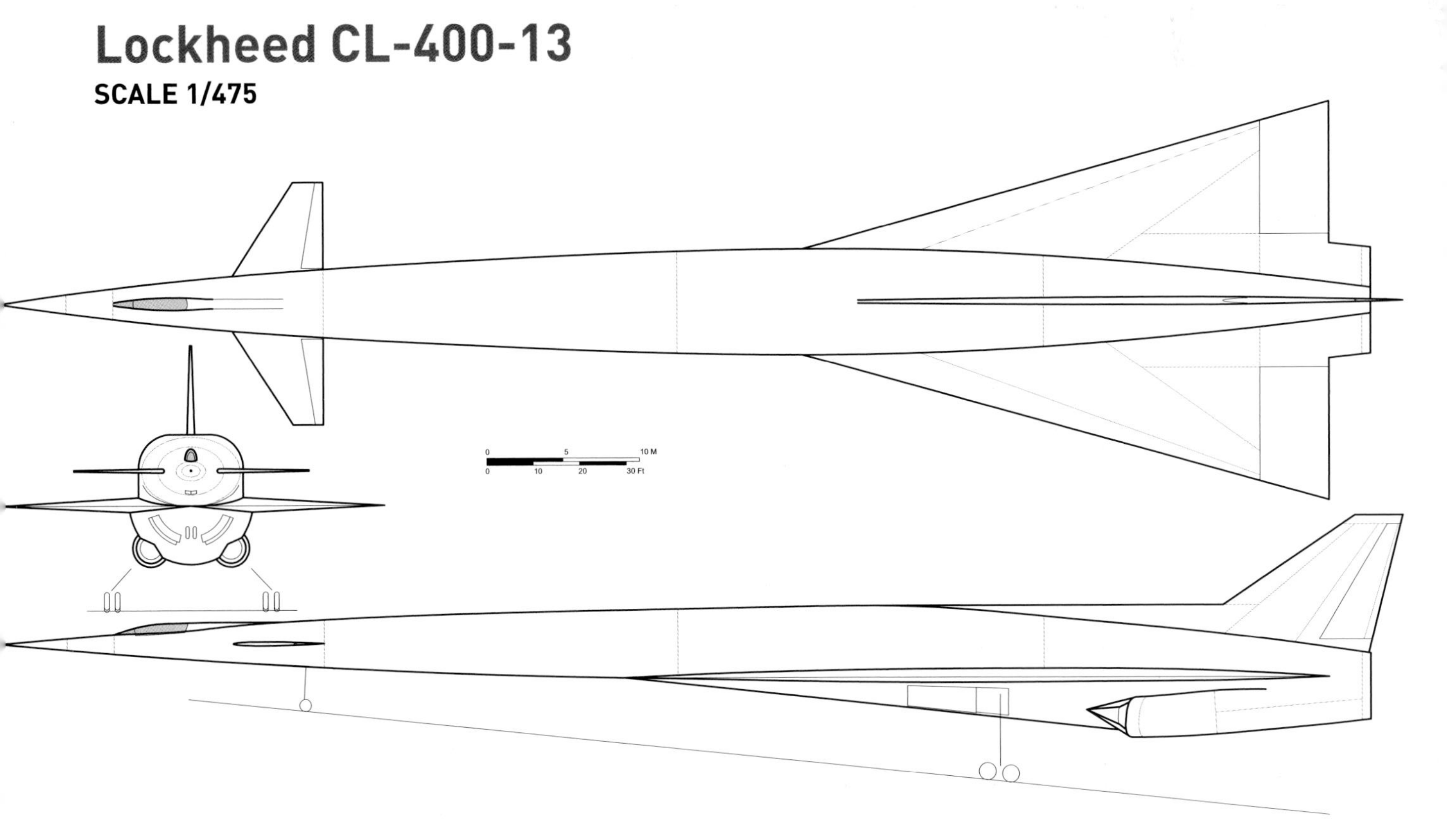

Lockheed CL-400-13
SCALE 1/475

Lockheed CL-400-14

SCALE 1/475

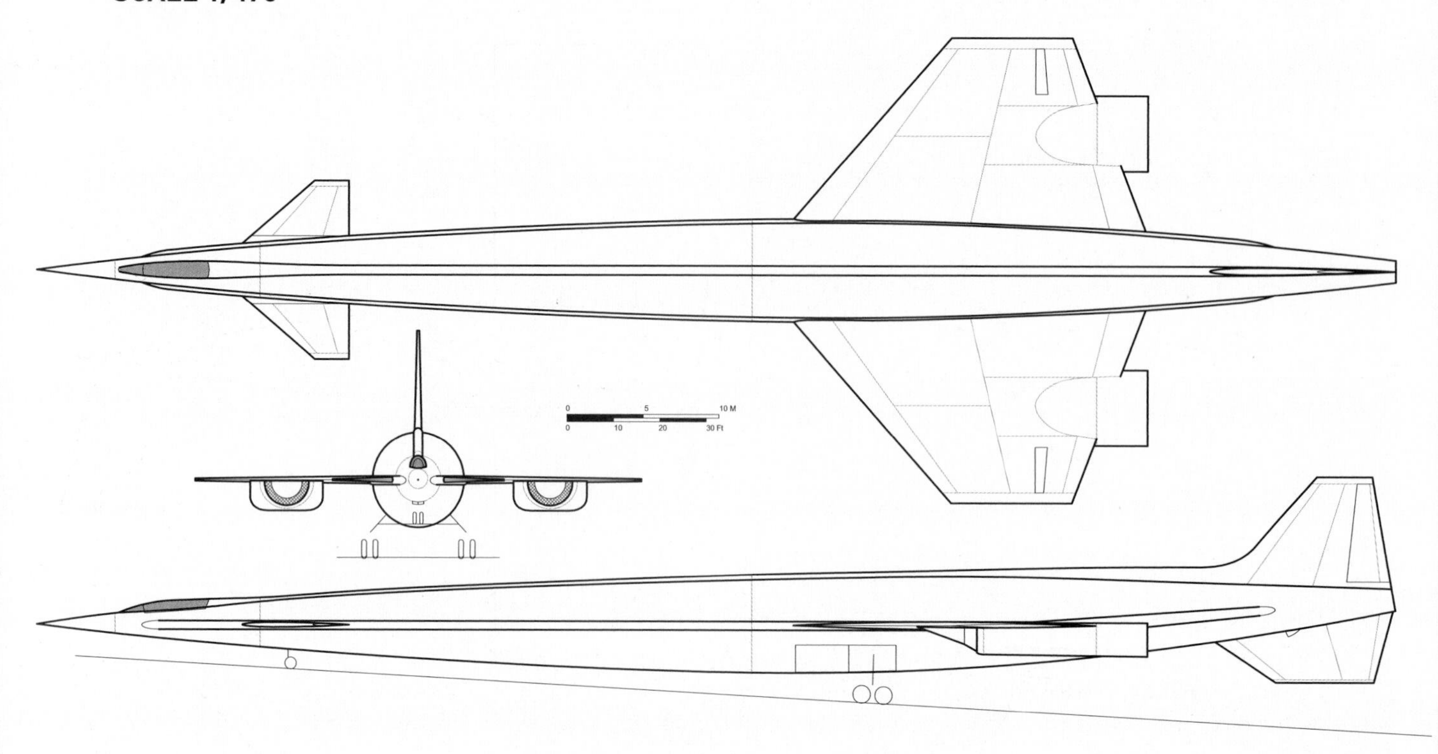

Lockheed CL-400-15 JP

SCALE 1/250

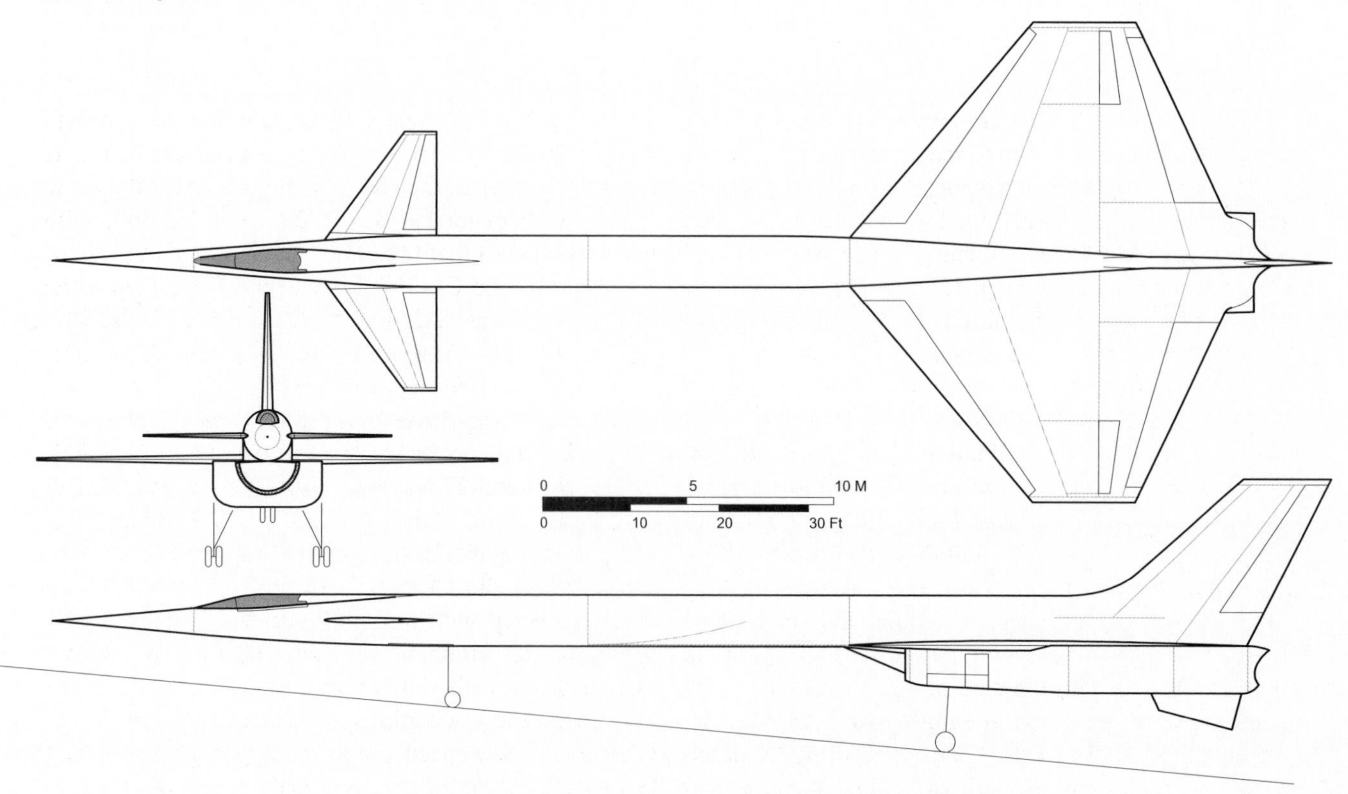

Bell System 118P

SCALE ~1/250, reconstructed from model photos

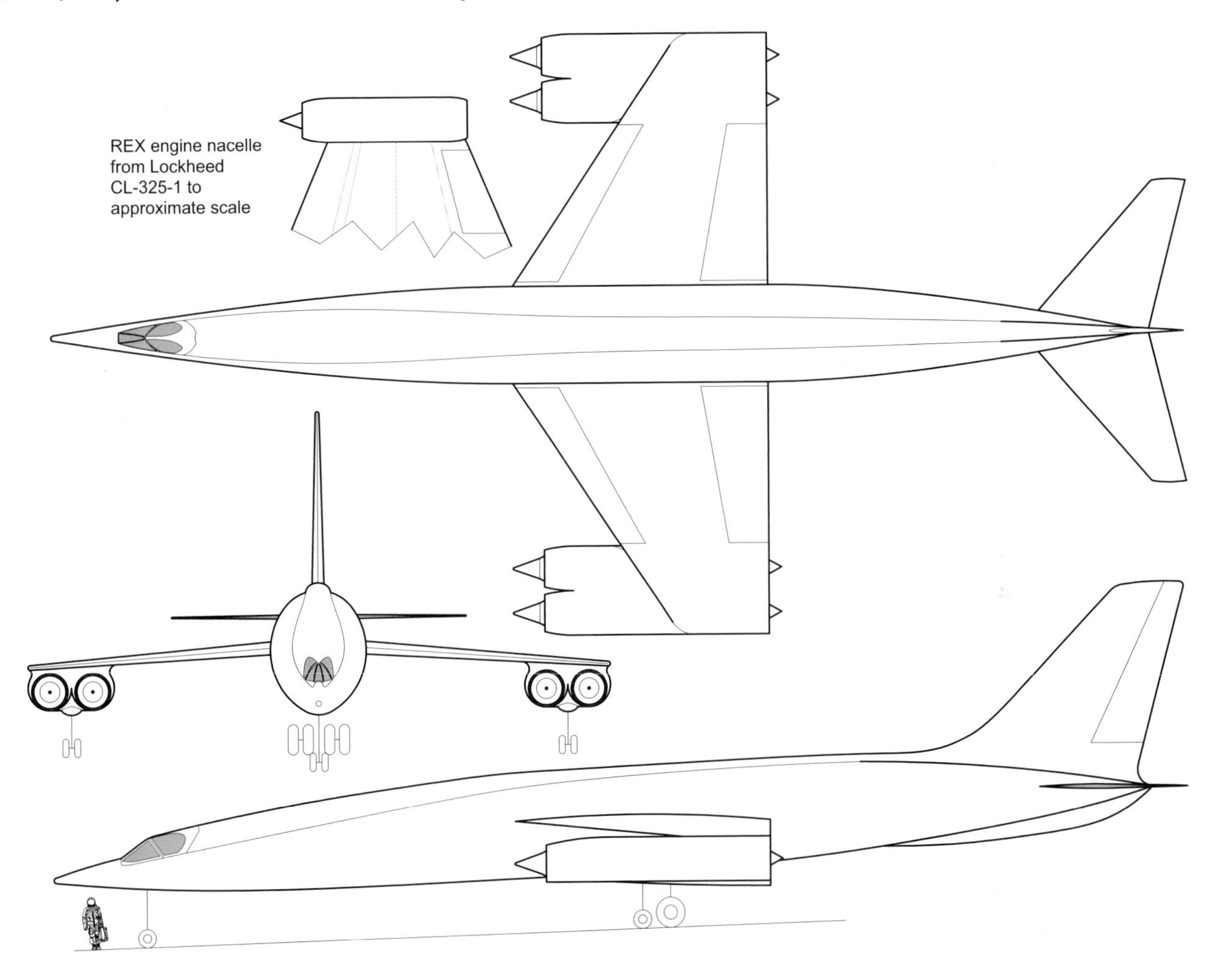

altitude to 100,000ft and an operational date of 1960. Other than that, the Air Force seems to have left options wide open. The known responses from the industry were quite varied. Northrop, for instance, submitted a large straight-winged six-engined subsonic vehicle, something of an overgrown U-2, for Phase II ½; for Phase III Northrop submitted an orbital spaceplane launched by a three-stage expendable booster rocket.

Two contractors are known to have produced designs akin to the CL-400 for WS 118P: Bell Aircraft and North American Aviation. The Bell design, frustratingly, is known only from a few photos of a pair of scale models. This design looks like it could easily have been part of the CL-400 design process, with a spindle-shaped bulbous body, barely-swept wings and four airbreathing engines in two wingtip pods. The diagram included here is a reconstruction based on those model photos; the size of the vehicle is a best approximation based on attempting to figure out the physical size of the 1/80 scale display model based on the details of the display case it was photographed with. The type of engines used are not known, though they are clearly airbreathers. If the reconstruction is accurately scaled, the Bell vehicle has about the same fuselage diameter as the CL-325 and CL-400, with engine nacelles that fall in between the size of those of the CL-325 and CL-400. It's currently unknown if the Bell aircraft was designed for Phase II ½ or Phase III.

The North American designs are far better documented. Two separate designs were reported... one to fulfill the Phase II ½ requirement (NAA model D265-26) and one for Phase III (NAA model D265-27). Phase III/D265-27 was, like Suntan, hydrogen fuelled, and used four Aerojet ATR-2040 air-turbo-rocket engines. The North American design was otherwise quite unlike the CL-400, featuring a flattened lifting fuselage filled with multi-lobe propellant tanks made of 6061-T6 aluminium. The wings were sharply swept deltas with substantial anhedral and fold-down wingtips. These wingtips would extend straight out to the sides at low speed giving the largest possible wing area. At high speed the wingtips would fold down to

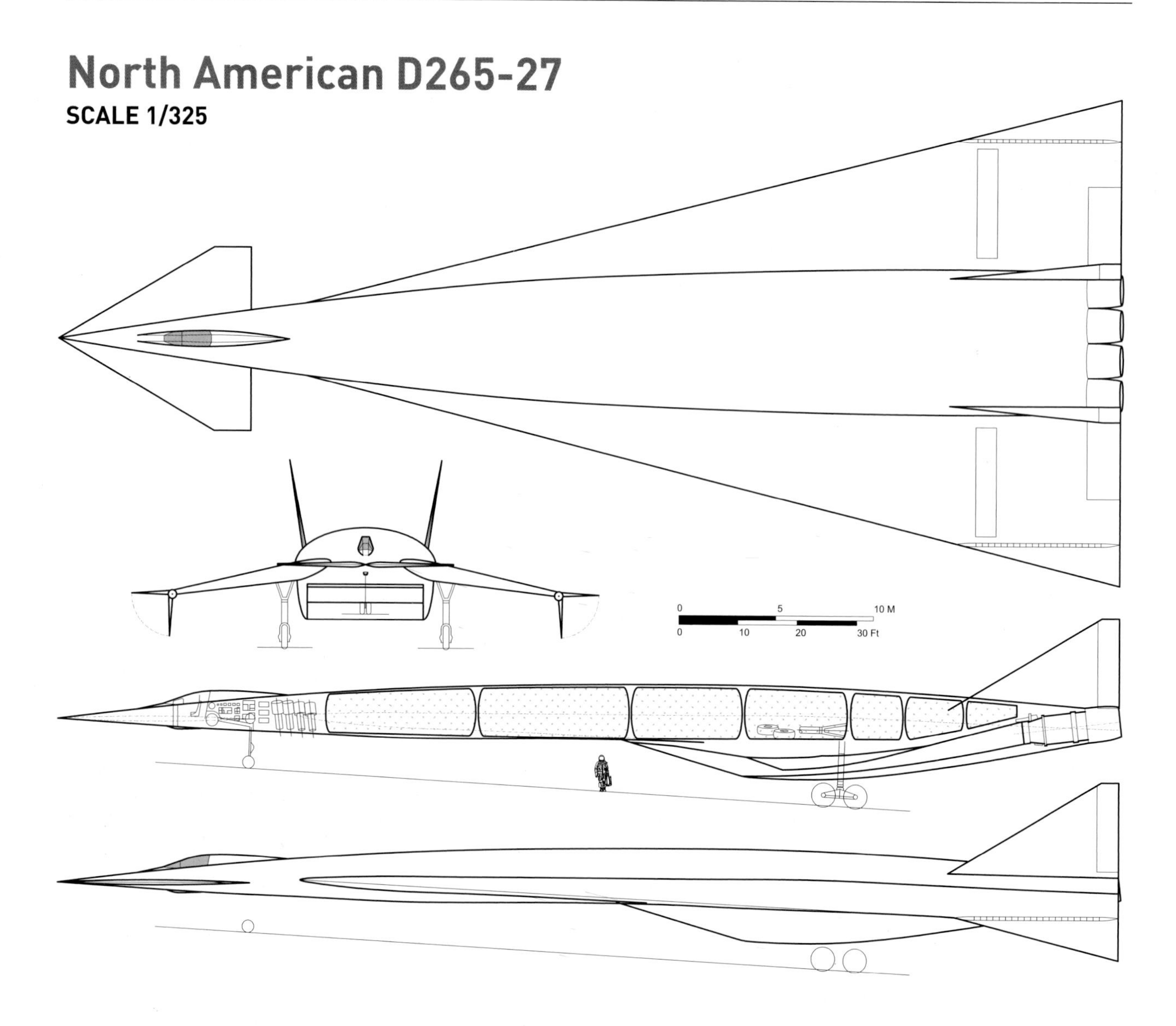

increase directional stability. Folding wingtips found their way onto the North American WS 110A entry in 1957, and eventually wound up being a successful design feature of the B-70… but for the B-70 the purpose was compression lift. The D265-27 had twin tails and delta canards mounted directly at the nose.

The four airbreathing engines were fed from a single two dimensional fixed ramp inlet on the underside. The inlet was as wide as the fuselage and would be fitted with a jettisonable cowl over the lower section during takeoff up to Mach 1.8. At that point the cowl would be jettisoned and, oddly, destroyed by explosive charges. AM350 stainless steel made up most of the structure, with brazed honeycombs forming much of the canards, outer wing panels and vertical stabilizers. The leading edges were to be Inconel X alloy for protection from the high temperatures expected from the high cruising speed. The forward fuselage would be skinned in titanium.

As of June 1956, the earliest possible operational capability would be 1963 due to engine development. In this case, the air turborocket from Aerojet, development of which had begun several years earlier. This engine was conceptually quite similar to the initial Rex-1 engine. Gas generators – essentially rocket engines – would drive turbines; the turbines would drive a compressor. Fuel rich exhaust from the gas generator would combust with compressed air to generate more power. Aerojet had already demonstrated an engine of this type, using hydrocarbon fuel; the D265-27 would require the development not only of a hydrogen-fuelled version, but a much larger one.

The NAA D265-27 was similar to the CL-400-13 in being a large and futuristic design, more at home in the 1980s as a single stage to orbit spaceplane. The North American D265-27 makes an interesting comparison with the North American B-70 and F-108 Rapier. These hypothetical stablemates were clearly from the same design family. It would have been quite a sight to see the Navaho, WS 118P, the B-70 and the F-108 – missile, recon, bomber, fighter – flying together.

CHAPTER 2 Competition from Convair and North American

The road to a dedicated American high-altitude supersonic reconnaissance aircraft was a long and winding one. For a number of years that road ran through the Convair Division of General Dynamics by way of the B-58 Hustler and related designs.

The first glimmers of what was to come emerged almost immediately after the Second World War as Convair began studying supersonic bombers. An early step on the road was the vast 'Generalized Bomber' (GEBO) study project at Convair which began in 1946. Reportedly 10,000 configurations for strategic bombers were studied – no doubt the vast majority being quite basic, or just slight variations on a theme. The final reports for GEBO were completed in 1949. This was followed up by the GEBO II effort, which expanded the field of study from conventional aircraft to multi-stage aircraft.

A number of different designs were examined over the next two years or so, but one concept that was looked at in some depth was a supersonic jet bomber equipped with a large bomb/fuel pod, carried aloft by a B-36 or B-36 derivative. At the time this made sense; turbojets of the day could push an aircraft past the sound barrier, but they'd burn through fuel incredibly quickly. Jet aircraft had woefully short ranges, and supersonic flight only made the problem even worse. But if a large, slow propeller-driven aircraft would carry the smaller supersonic bomber most of the way to the target (or at least across an ocean), it could then release the bomber to fly to the target under its own power at high speed.

One such design, dating from June 1951, is shown here. This is recognizable as an early ancestor of the B-58, though many design features would change drastically. The bomb/fuel pod is substantially larger than the fuselage of the aircraft, and contained not only the bomb and fuel, but also a turbojet and radar. Once separated from the aircraft, it would continue on under power for a short distance as a standoff weapon.

The bomber itself was equipped with three turbojets, one in the tail and two in individual underwing pods. Unlike the podded engines of the B-58, the two podded engines in this design were sacrificial: after turning tail and dashing out of enemy territory, the bomber would drop the engines and continue home at slower speed with the single remaining engine. Consequently, a single flight of this aircraft would result in the loss of a large amount of expensive hardware.

For the carrier aircraft, the GEBO II design shown here would use not the B-36, but the B-36G, later redesignated the B-60. The B-60 added swept wings to the B-36 and removed the piston engines, replacing them with four individually podded turboprops. The GEBO II bomber would take off equipped with eight engines but return with only one, having separated from four of them and thrown away three more.

Following the GEBO II studies, Convair refined its work into a design under USAF Weapons System MX-1626. First shown to the USAF in January of 1951, it was generally similar to the earlier studies: a relatively small aircraft with highly swept delta wings, an underslung bomb/fuel pod larger than the aircraft's fuselage, and carried to altitude/range underneath a swept-wing turboprop-equipped B-36 derivative. The pod contained one turbojet, the fuel for the outbound leg and either the nuclear weapon or reconnaissance equipment, along with navigation radar in the nose. The pod also had three tail fins for stabilization; the lower fin could fold to provide ground clearance.

The aircraft had two engines mounted roughly at mid-span in long nacelles. Mounting them this far out put them beyond the boundaries of the carrier aircraft's bomb bay. Perhaps oddly, the pod was meant to fall ballistically after separation from the aircraft; it does not seem that it was intended to use its own turbojet to increase delivery range. The upper surface of the pod was flat, mating to a flat undersurface of the aircraft fuselage. This would reduce drag while the pod was attached to the aircraft.

The wings themselves a 65° delta with a simple symmetrical airfoil. As Convair continued to develop the delta wing from the XP-92 through the F-102, F-106 and B-58, the wing would grow a noticeable leading edge camber. The aircraft did not have the area ruling that would give the F-106 its distinctive 'wasp

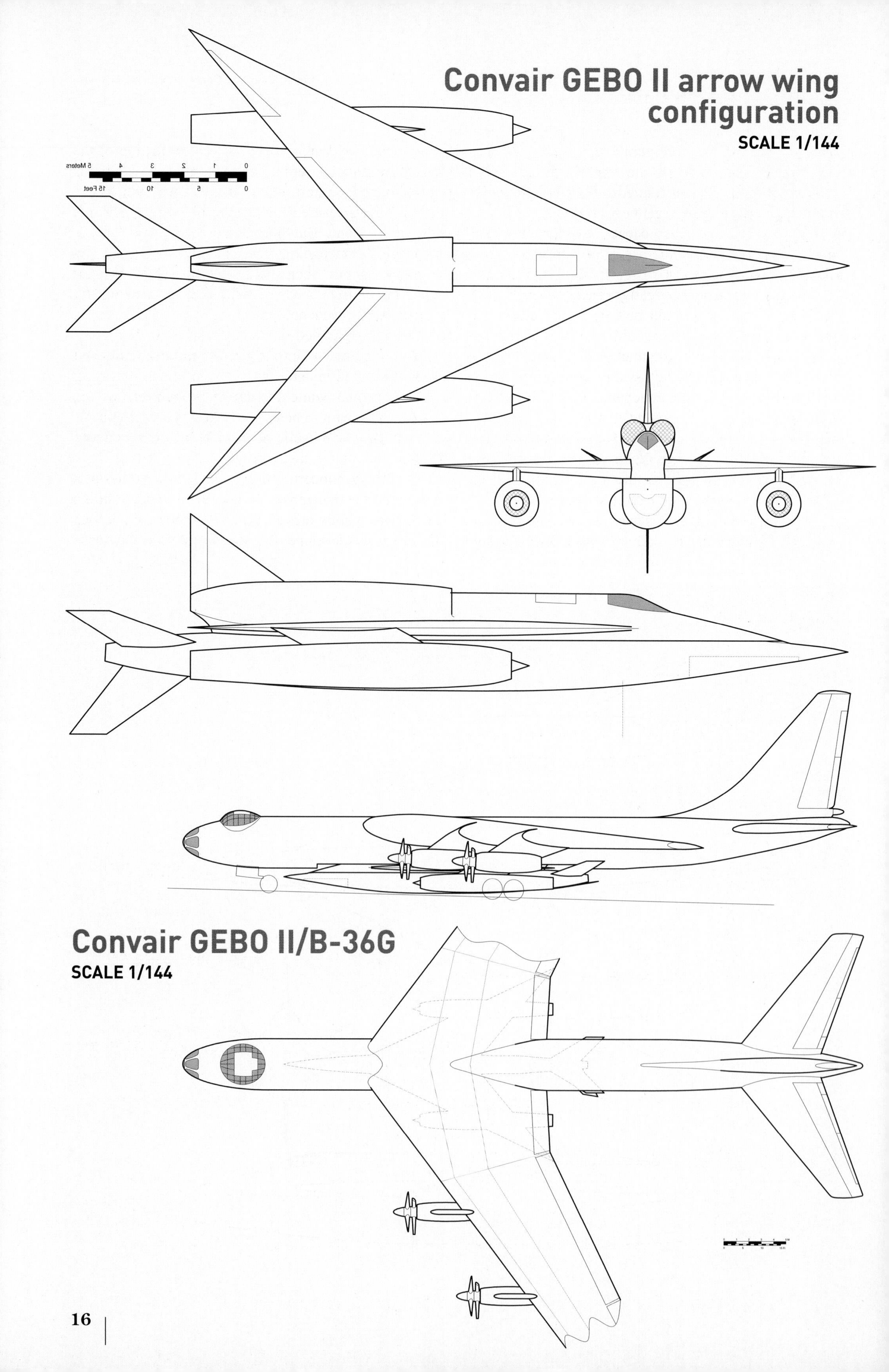
Convair GEBO II arrow wing configuration
SCALE 1/144
Convair GEBO II/B-36G
SCALE 1/144

waist'. The overall configuration of the MX-1626 was quite reminiscent of the later Lockheed A-12, though substantially less refined and elegant. It was an interesting case of convergent evolution in action.

The MX-1626 was also capable of launching itself from a runway without the use of a B-36/B-60 carrier aircraft, although this would come with a substantial performance penalty and greatly reduced range. In order to do it, the aircraft would be fitted with jettisonable tricycle landing gear which would fall away immediately after liftoff. Whether self-launched or dropped from a carrier plane, the MX-1626 was incapable of landing with the belly pod still attached. Once airborne, the aircraft would need to expend the pod, destroying the expensive equipment on board. This would necessarily make training and practice flights rather expensive.

And for all the effort, the performance of the MX-1626 was surprisingly modest, with a maximum speed of only Mach 1.5. This may have been respectable for 1951, but within a relatively short time it would seem positively lacklustre. This was something of the end of an era in conceptual aircraft design. Up until this point 'staged' aircraft, carried to distance by a larger, slower aircraft, were seen as the only way to achieve certain missions. But during the early 1950s air-to-air refuelling technology and techniques improved by leaps and bounds. Soon it was clear that achieving great range via in-flight tanking made much more sense. But staged aircraft were not quite done.

The MX-1626 led directly to the B-58 'Hustler'. This is of course a famous aircraft and does not need to be described in detail here. But what is interesting to note is that while the direct progenitors of the B-58 were meant to be carried aloft by larger, slower aircraft, the B-58 itself ended up being proposed as a carrier for smaller, faster aircraft.

The thing modern amateur aviation enthusiasts may not fully realize is that the pace of change in the 1950s was wholly different than it is today. Back then aircraft were developed on schedules often measured

Convair MX-1626

SCALE 1/144

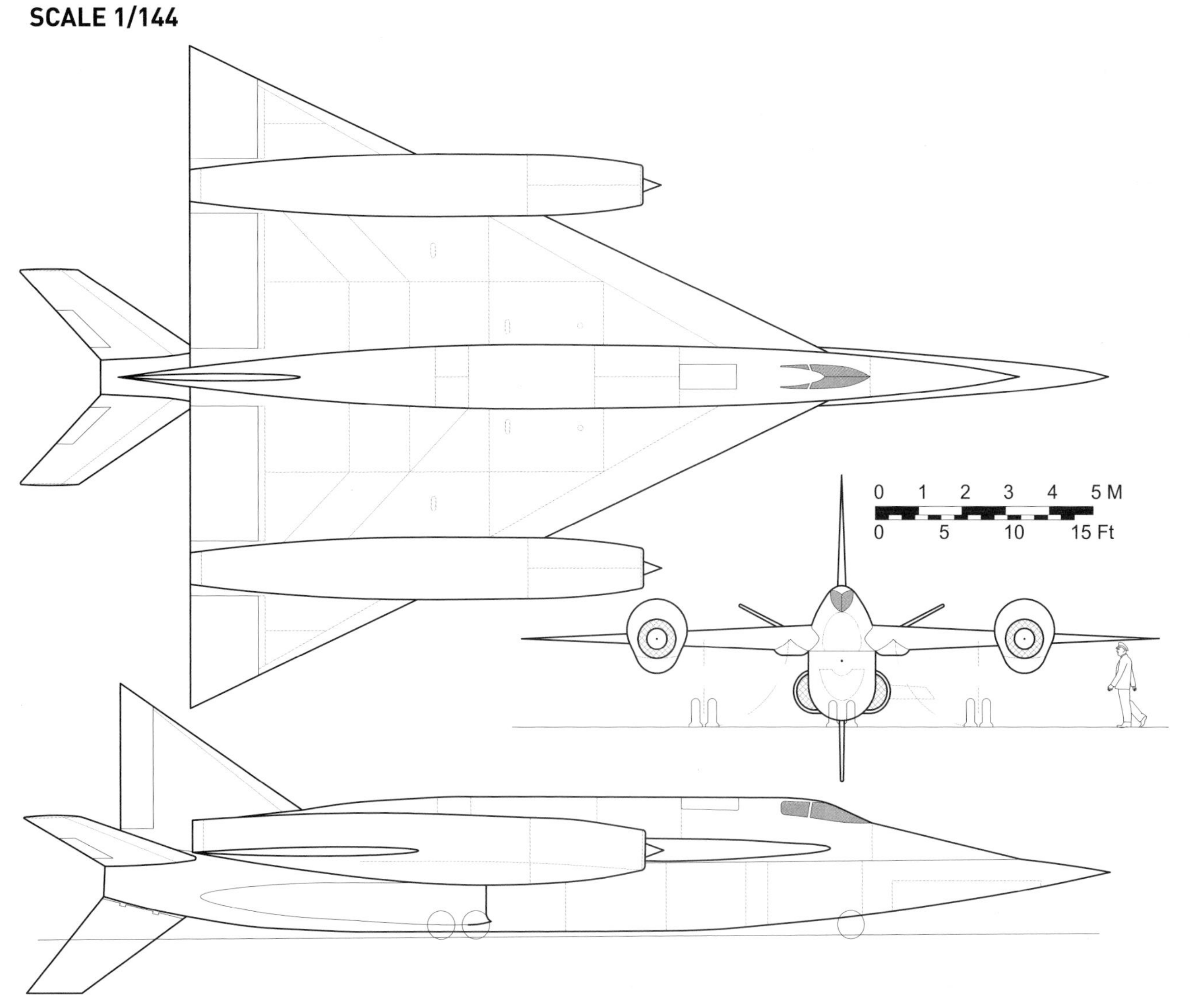

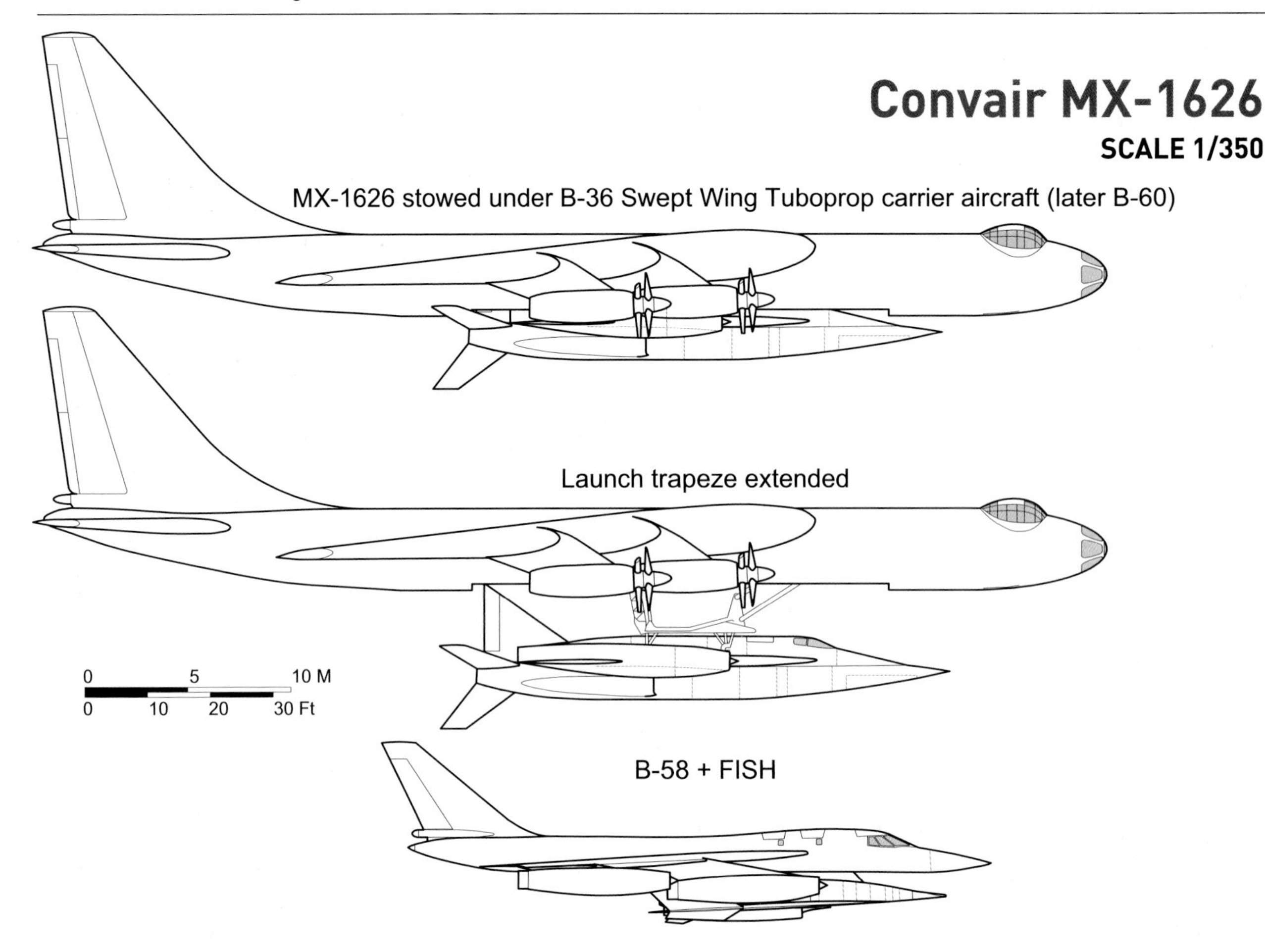

in months, while today's most advanced fighters and jetliners are often designs that were selected for development decades ago (the Lockheed design for the F-22 was selected for development in 1986, some 34 years ago at the time of writing; the Lockheed F-35 was selected in 1997, 23 years ago; the Boeing 787 jetliner was announced in 2003, 17 years ago) and will trudge on in service for decades to come. In the heady days of the 1950s, advanced aircraft were often obsolete before they flew. And so it was common practice for the designers of the latest and greatest aircraft to start drafting improved versions, derivatives and follow-ons before the ink was dry on the baseline design drawings.

So during 1955, long before the B-58 would make its first flight (in late 1956), the engineers and designers at Convair's Fort Worth, Texas, facility were already looking at ways to greatly improve upon its performance. Potential upgrades included replacing the engines, stretching the fuselage, refining the aerodynamics and so on. Over the next few years numerous studies were made to increase speed to Mach 4. Advanced fuels, turboramjets, auxiliary ramjets and rocket engines all found their way into the designs, yet all were of somewhat dubious value due to uncertainties regarding cost, performance and materials. In late 1957, Convair began studying another approach to expanding the performance envelope of the B-58: a small parasite.

The initial concept called for a single manned vehicle to be carried under the centreline of a stretched Hustler in the position normally occupied by the bomb/fuel pod. As initially conceived, the parasite would be ramjet powered and designed to cruise at Mach 5… but there was a problem. With any practical single vehicle, the B-58's main landing gear would be blocked by the parasite's wings. Soon a solution was found by splitting the parasite into two components. The parasite, dubbed the 'Super Hustler', had an expendable first stage equipped with two ramjet engines and a ramjet-powered cruise stage with a single pilot. Both stages had their own wings; the result was that the B-58s main landing gear would have clearance to retract between the two sets of wings. By early 1958 the initial concept had crystalized into a detailed concept with a crew of two.

The booster had two side-by-side Marquardt ramjet engines in the rear of a stubby fuselage, each fed by a rhomboidal external-internal compression inlet on either side of the fuselage. Swept shoulder-mounted wings featured sizable downward-angled wingtip fins. The sharply pointed nose of the booster fuselage contained a single nuclear warhead, the underside of which mated to the upper rear fuselage of the manned cruise vehicle. The cruise vehicle had a flat underside and sharply swept arrow-shaped faceted wings, with similarly shaped vertical fins at about mid span. The

Convair Super Hustler Configuration 121

SCALE 1/120

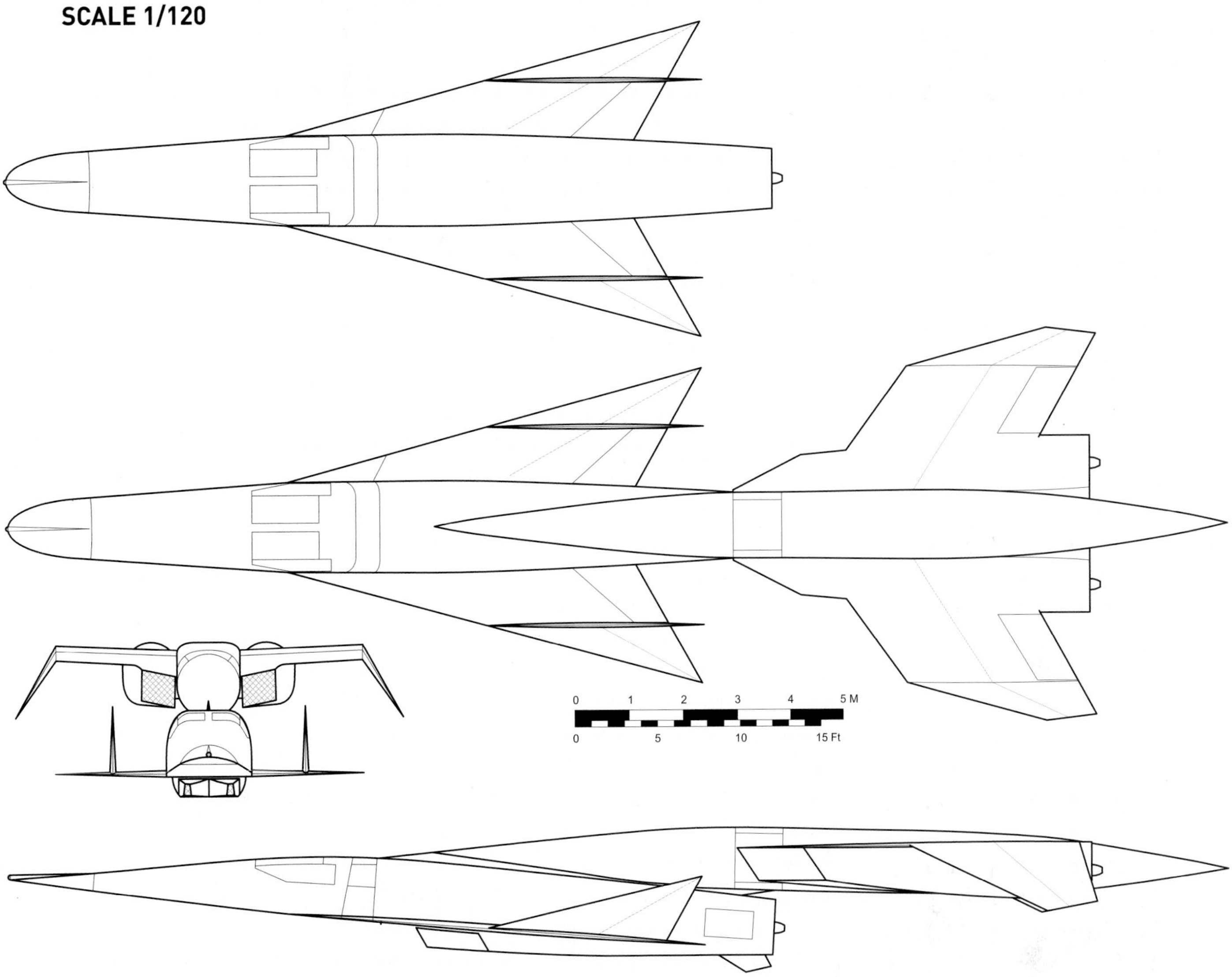

0 1 2 3 4 5 M

0 5 10 15 Ft

Convair Super Hustler Configuration 124

SCALE 1/120

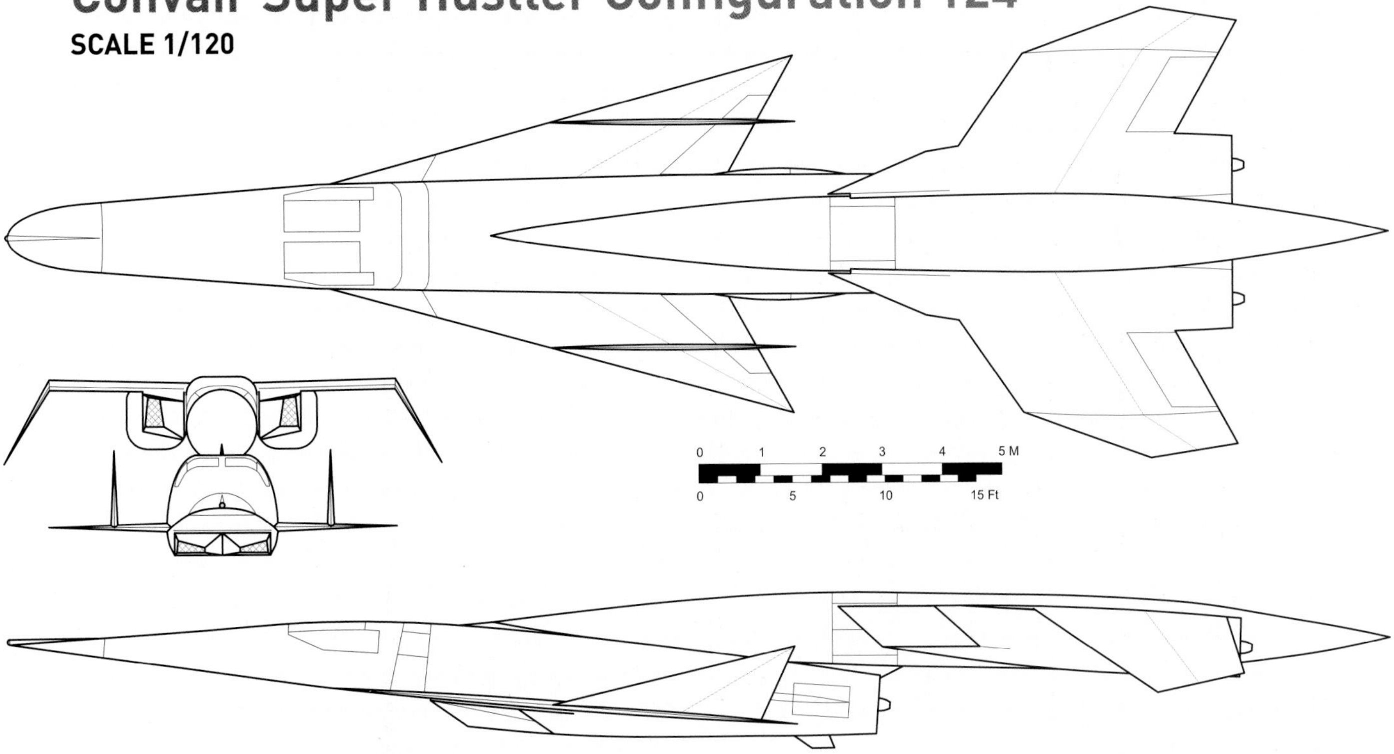

0 1 2 3 4 5 M

0 5 10 15 Ft

baseline Super Hustler had a single ramjet engine in the cruise stage; an alternate design had two side-by-side ramjet engines. Both designs used an underslung inlet for the ramjet, with an auxiliary small turbojet used for landing.

The B-58 would carry the Super Hustler to a speed of Mach 2 and an altitude of somewhere under 40,000ft. The Super Hustler would be released and would fire up the ramjet engines in the booster, accelerating to a cruise speed of Mach 4 and an altitude of 73,000ft. Speed would stay at Mach 4, but altitude would gradually increase as fuel was burned off. A standard mission would see the booster separate from the cruise vehicle around 2000 miles from B-58 launch at an altitude of 80,000ft. After separation from the carrier, the warhead would separate from the booster stage and would plummet towards the target, with an expected circular error probability of 1410ft. The warhead would deploy an aft skirt for stabilization. A skirtless lower-drag design would reduce CEP to 1100ft. The manned stage would continue to cruise at Mach 4, climbing to a maximum of 91,000ft, before eventually landing some 4290 nautical miles from the point of B-58 separation.

The Super Hustler manned stage was a small, compact vehicle, yet it could attain high speed, high altitude and substantial range. This was due to the fact that it was designed to stay at on-design conditions. A ramjet engine is generally considered to have poor fuel economy, but this is due to the fact that ramjets only have poor fuel consumption when flown at anything other than their optimum design condition.

Acceleration from low speed and climbing from low altitude will drain a ramjet's fuel tank in no time flat. But the Super Hustler manned stage had the advantage of being boosted to its design condition, and could thus operate at maximum efficiency. This would come in handy for a proposed alternate use of the Super Hustler as a dedicated reconnaissance vehicle. With the bomb replaced with fuel in the booster, the whole vehicle could attain greater range. With a high resolution camera system in the nose along with several radar systems (a 360° scanning radar with a range of 180 nautical miles and a high resolution X-band ground mapping radar with a resolution of 200ft at a distance of 20 nautical miles), the vehicle could reach somewhere around 6600 nautical miles range after launch from the B-58.

The nose of the manned stage was required to fold down and back to provide clearance for the B-58's landing gear. In addition, the Super Hustler was low and flat and did not have a raised cockpit canopy. Given the relatively low lift-to-drag of the vehicle, it would have needed to land at a high angle of attack and this would have precluded any realistic possibility of the pilot having anything remotely resembling decent forward vision during landing. The aircraft would therefore be equipped with a television camera in the extreme nose and the entire nose from just aft of the cockpit could tip downwards 20-degrees. The end of the nose would also tip upwards to provide some clearance. And a series of doors that covered transparencies over the cockpit would open up, allowing the crew to actually see outside. Aerodynamically, the manned stage would have been an interesting mess as it came in for a landing on a nosewheel and two tail skids.

Furthermore, the fact that the forward fuselage could angle away from the rest of the aircraft created a useful opportunity. In the event that the aircraft was stricken, the entire forward fuselage could break away, accelerated forward by a solid rocket motor. It would deploy stabilizing flaps until it had slowed below Mach 1. At 30,000ft and Mach 0.9 it would release a drogue chute to slow it further; two 60ft diameter recovery chutes would deploy at 25,000ft and Mach 0.25 and lower it safely to the surface.

From Super Hustler to FISH

The use of an expendable booster stage would make the Super Hustler a complex and expensive aircraft to train with. Apparently parachute recovery was proposed for use during training, but that would still be expensive. For peacetime reconnaissance missions, the booster stage would likely come down in Soviet or Chinese territory, an issue problematic on several levels. During wartime it would hardly matter if vehicles started crashing down out of the sky, but in peacetime it would at the very least alert the enemy as well as providing them with indisputable concrete evidence of overflights. So the idea of a single stage parasite returned: without a booster to expend, a reconnaissance aircraft could overfly enemy territory in peacetime.

Additionally, the Super Hustler, while small, was detectable. There were few practical provisions for radar or infra-red stealth in the original design, and reconnaissance was merely an afterthought. To be a truly useful reconnaissance aircraft it would need to be able to fly secretly. To that end, the Super Hustler was redesigned into the Special Purpose Super Hustler. The initial concept was the 'Minimum Change Configuration', where the wings and vertical fins were redesigned; the straight leading edges were turned into convex curves. This was done to reduce radar returns from specific directions. Two slightly larger ramjets were employed, the fuselage was stretched and the booster stage was eliminated entirely. This was apparently a minor study, but it led to important new directions.

The Special Purpose Super Hustler evolved from

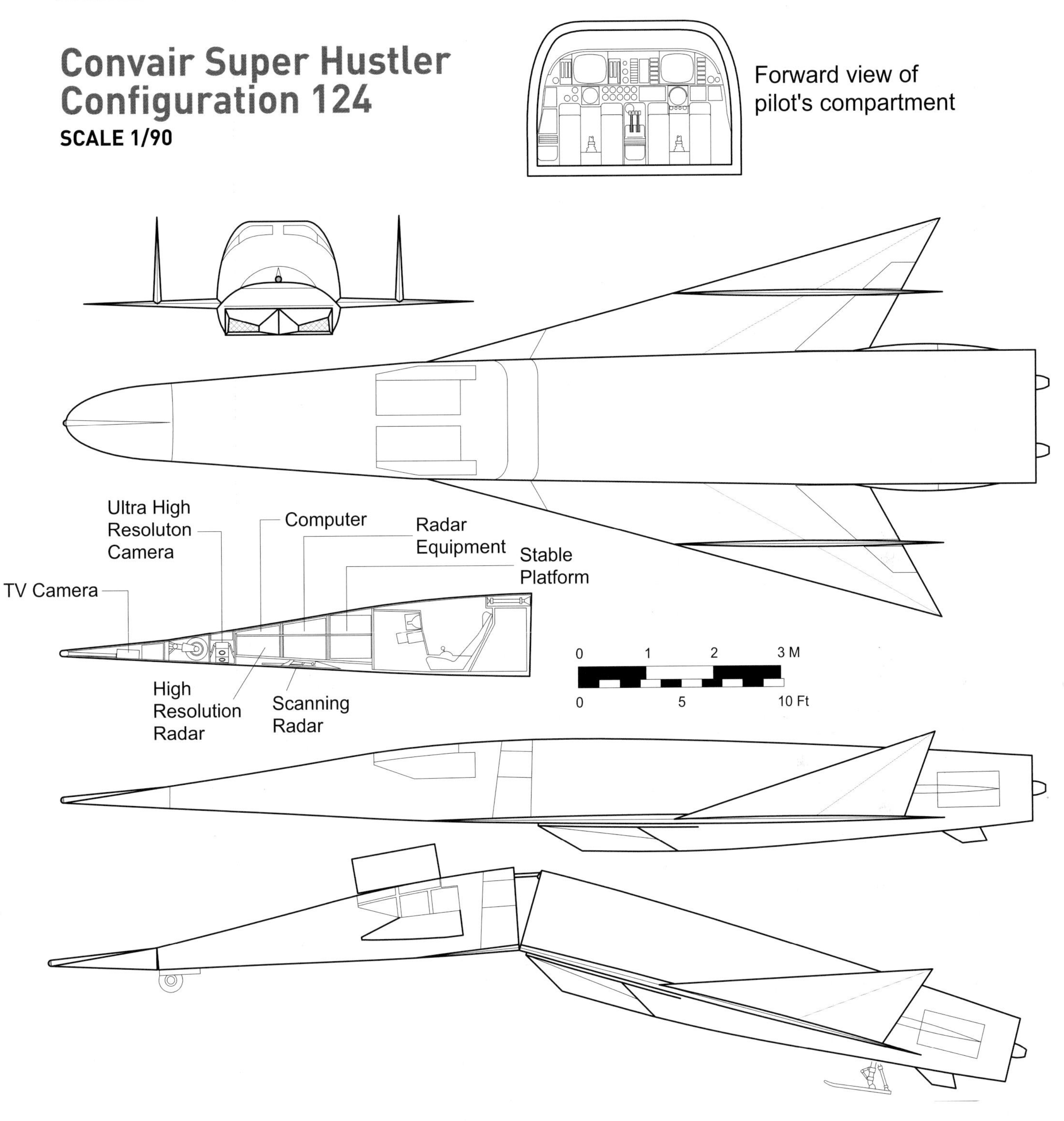

the Minimum Change Configuration. The fuselage grew somewhat 'fatter' and the wings gained substantial surface area while retaining the unusual curved leading edges, and added curves to the trailing edges. This configuration was preferred and was dubbed Configuration 220.

It was at this time, November 1958, that the name of the design transitioned from Super Hustler to FISH: First Invisible Super Hustler. Super Hustler had been optimized for performance, but FISH was designed for low radar cross section. Super Hustler would have not only been a sizable radar return in the sky, it would have dropped airplane-sized pieces of hardware onto enemy territory. FISH, on the other hand, was intended to cross Soviet skies unnoticed. It would do this in part due to its relatively small size, high speed and great altitude, but also due to geometry and construction that would reduce its radar return. FISH was still carried aloft by the B-58, but it was now a single vehicle – no expendable booster stage. Equally importantly, it was designed from the outset for low radar cross section.

The November 1958 design was derived from the manned stage of Super Hustler, but it was nearly

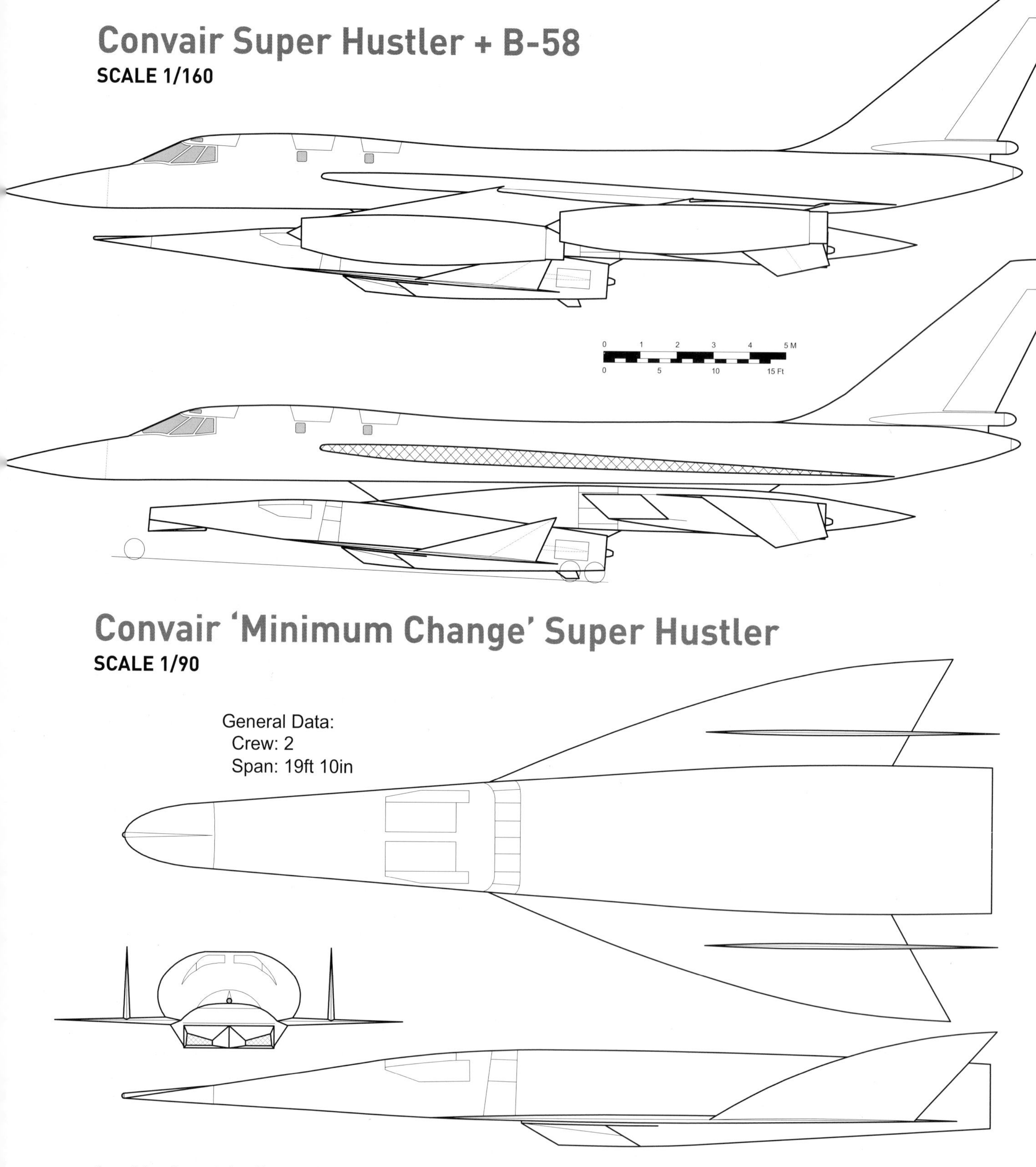

devoid of straight lines; they were replaced with gentle curves to scatter radar. Right angle corners were reduced as much as possible. The vertical fins required extensions through to the underside of the wings; these new ventral fins were tipped substantially inboard, eliminating the 'corner reflector' effect that would send a radar beam directly back to the emitter. The main portions of the vertical stabilizers above the wings were still perpendicular to the wings. Given that the aircraft was designed to cruise far above any sources of radar, this was a reasonable design decision. The underside of the craft was mostly flat with the exceptions of the ventral fins and the ramjet inlets. Large models of the

FISH were built and were put through an extensive series of RCS measurement tests to help refine the design for minimum observability.

Along with the increased area and curved planform, the wing was now fitted with radar absorbent 'teeth' on the leading and trailing edges. Wedge-shaped dielectric inserts (made from the glass-ceramic material Corning Pyroceram, to take the 915°F cruise temperature) formed the edges and would absorb incoming radar energy rather than reflecting it back. The leading edge mechanical structure was made of René 41, a nickel 'super alloy' used for the Mercury capsule exterior shingles due to its temperature resistant properties. During high-speed cruise the metal structure would thermally expand more than the ceramic inserts, which were designed to slip back into their triangular slots as the metal airframe expanded around them.

As with the Super Hustler, the nose of the FISH Configuration 220 was designed to fold down and back to provide clearance for the B-58's nose gear and included the TV camera. The whole nose of the FISH featured the 'drooping' and jettison features of the original Super Hustler. But since it was now a single-crew vehicle, it had only half as many insulated doors over the canopy. What was once the reconnaissance system officer's half of the cockpit was now taken up by a fuel tank. In order to provide clearance for the B-58's main landing gear, the relatively short FISH was mounted as far forward as practical.

FISH was a dedicated recon vehicle with no provision made for carrying armament. It would be launched from the B-58 at Mach 2.2 and would accelerate under ramjet power to Mach 4 and climb to 90,000ft. The FISH would cruise at that speed and altitude for about 50 minutes, turn around and fly home using Marquardt MA24E ramjets burning a boron-based high energy fuel. The return leg of the trip would see the FISH climb to short of 100,000ft altitude. At the end of the mission, a single Pratt & Whitney JT12 turbojet would provide subsonic flight and powered landing capability.

Convair wanted to base the FISH at Carswell Air Force Base in Texas, near the Convair plant in Fort Worth. This would have been convenient on technical, logistical and security grounds. The FISH itself, at the end of each mission, would land at either an overseas US air base, or perhaps even on a US Navy aircraft carrier. In either case, returning the craft to Texas would be a bit of a chore given that it was not capable of self-ferrying. It would need to be carried under a B-58 for a return flight. This would not be possible for a carrier-landed craft until the carrier pulled into a port with a convenient nearby US air base.

The FISH Configuration 220 was a much more elegant and refined vehicle than the Super Hustler. That is especially apparent in the inlets: those of the Super Hustler were flat-walled and angular, while the FISH inlet had an external ramp of compound curvature.

A minor variant of the FISH was briefly studied when it was still referred to as the 'Special Purpose Super Hustler'. An 'unstaged' design was looked at that was capable of launching itself with no need of a B-58. In order to accomplish this, the ramjets were removed from the fuselage and put into individual cylindrical nacelles at the wingtips. The standard Configuration 220 had a single non-afterburning P&W JT12 turbojet for use at landing; the 'unstaged' vehicle had two of them in an extended rear fuselage, with afterburners. The JT12 was a fairly small engine, so even with afterburning getting the aircraft up to Mach 3 seems a bit optimistic.

By June 1959 the FISH configuration had changed noticeably. On Configuration 234, the cockpit was no longer fully submerged. Giving the pilot direct forward vision meant the television camera in the nose was no longer needed. In order to fit closely against the underside of the carrier aircraft, the cockpit was offset to the port side. The ability to jettison the entire forward fuselage was removed and the pilot was given a B-58-style ejection capsule instead.

The vertical stabilizers were moved from the wings and located on the side of the rear fuselage, with the ventral fins entirely deleted. And instead of a single turbojet submerged within the rear fuselage for landing, two deployable turbojets were located within the sides of the fuselage just aft of the cockpit. At last the vehicle also had wheeled main landing gear rather than skids.

Configuration 234 was noticeably heavier than Configuration 220. Both the empty weight and fuel load increased, with range dropping by some 250 miles. Perhaps worse, the added weight meant that the standard B-58 was no longer able to effectively carry it. So the B-58 would have to be modified.

The B-58B carrier aircraft proposal made three main changes. First, the J79-5 turbojets would be replaced with more powerful J79-9 engines, the fuselage would be stretched 5ft to make room for additional fuel and the vertical stabilizer would be enlarged to counter the destabilizing effect of the new longer fuselage. It was proposed that the B-58B would be created by modifying airframes currently being built as B-58As. The FISH was still located ahead of the B-58's main landing gear, but the fuselage stretch meant that the nose of the parasite aircraft no longer had to fold back for clearance. All of the former variable geometry features of the nose and forward fuselage of the FISH were now deleted.

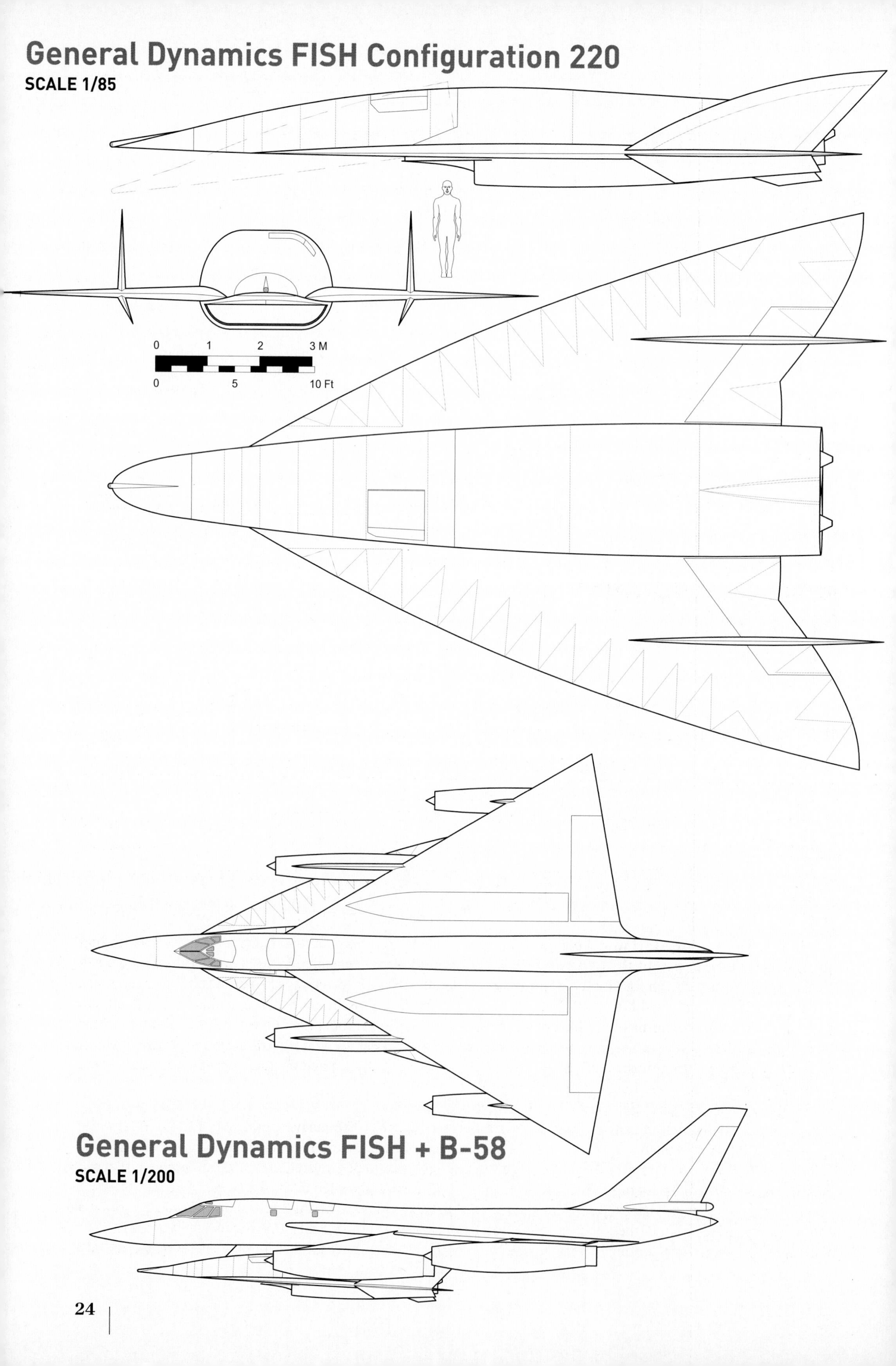
General Dynamics FISH Configuration 220
SCALE 1/85
0
1
2
3 M
0
5
10 Ft
General Dynamics FISH + B-58
SCALE 1/200

General Dynamics FISH Configuration 234

SCALE 1/85

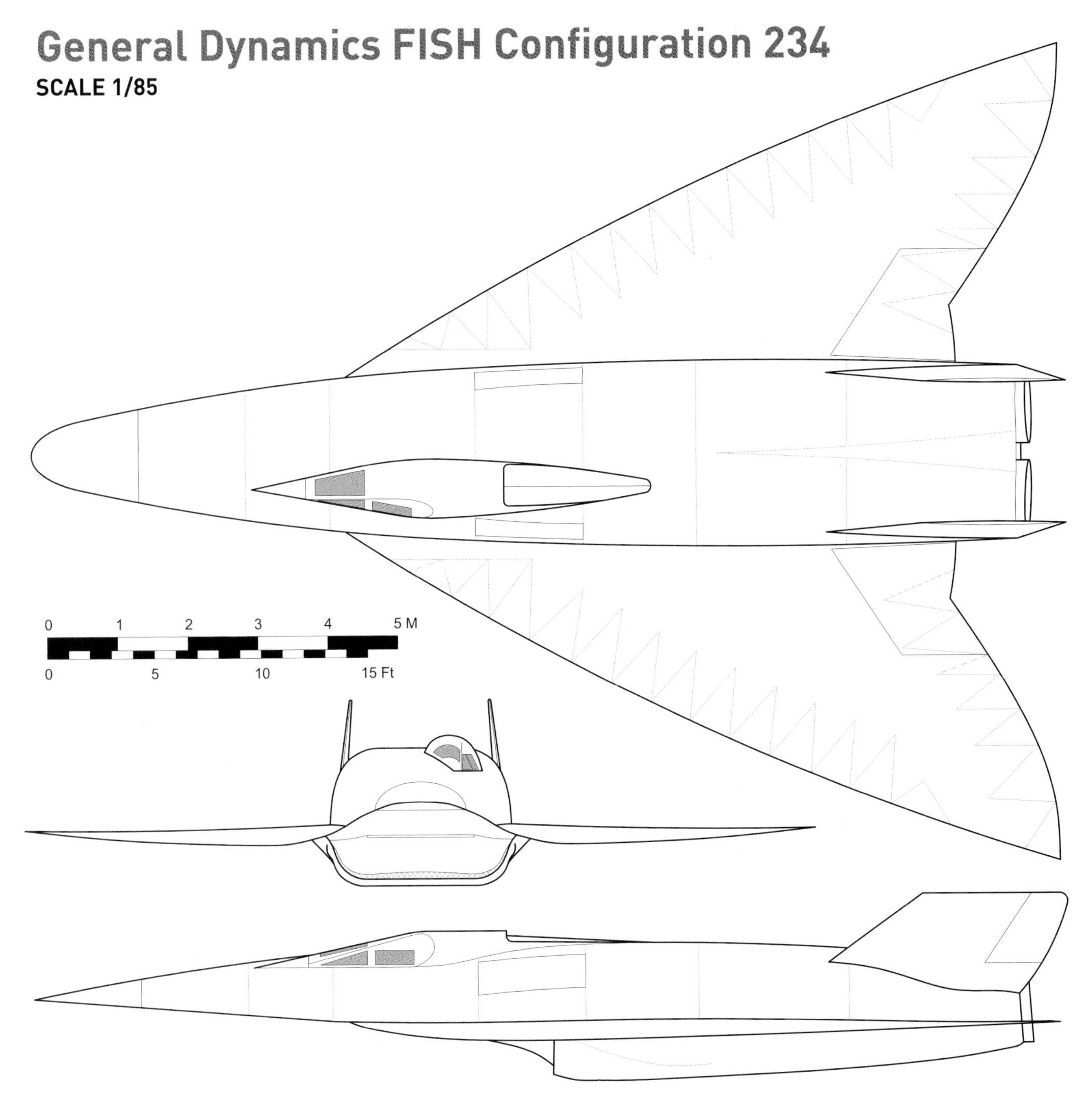

The total range of the FISH was dependent not only on the independent range of the parasite aircraft but on the range of the carrier as well. Without inflight refuelling, the carrier could transport it only 900 nautical miles from its airbase to the launch point. But with a single inflight refuelling that range would increase to 2100 nautical miles; with inflight refuelling twice, 4200 nautical miles. Launching airfields such as London, England and Fairbanks, Alaska were considered. Potential landing sites included Okinawa, Japan and Karachi, Pakistan.

The B-58B carrier aircraft was related to the B-58B that Convair was attempting to sell to the Air Force. The two planes shared the fuselage stretch and new engines, though the other B-58B was a dedicated bomber. The bomber version included small canards initially, which turned into leading edge root extensions. But the B-58B failed to sell to the Air Force as a new bomber, with the consequence that funds would not become available to make them in small numbers to carry parasite reconnaissance aircraft. And without its carrier aircraft, FISH was left unable to launch and needed to be redesigned.

From FISH to KINGFISH

Along with the unavailability of the B-58B carrier aircraft came new requirements: the aircraft was now supposed to be powered solely by turbojets, not ramjets. These two facts led to the FISH evolving into a single-stage vehicle. With this major change in design came a new name: KINGFISH. Convair had only a month, from June 1959 to July 1959, to completely revise the aircraft.

General Dynamics FISH Configuration 234 + B-58B

CALE 1/200

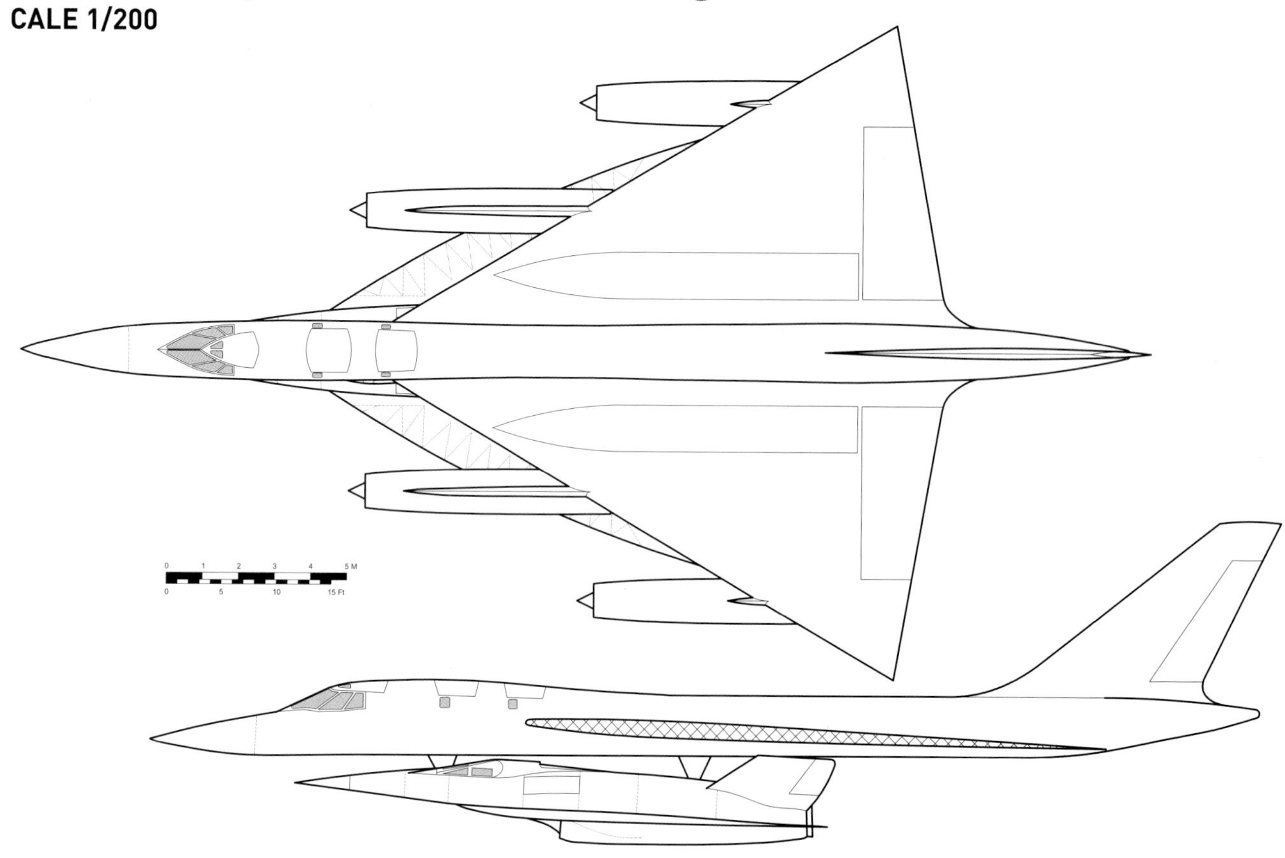

The first major step in the new evolution was Configuration 238 from July 1959, dubbed 'Kingfish Lower' due to the inlet being on the lower side of the aircraft. Configuration 238 was shaped much like Configuration 234, just scaled up by about 50%. It even retained the raised cockpit offset to port. The trailing edge of the wing differs from illustration to illustration, sometimes shown as an unbroken arc, sometimes depicted with a cutout just aft of the engines. The inlet was moved much further forward than on the Configuration 234, doubtless due to the turbojets in the fuselage being relatively much longer than the fairly stubby Marquardt ramjets.

Configuration 242-A came along at basically the same time as Configuration 238 and revised the design for improved radar cross section. The inlet was removed from the underside of the craft and split into two, with one inlet on either side of the fuselage. The underside was now smooth and unbroken. The inlets were inverted 2D ramps; these would have done an excellent job of shielding the interior of the inlets from surface-based radar systems, at the expense of pressure recovery issues at low speeds and high angles of attack.

In keeping with the 'FISH' theme, Configuration 242-A was dubbed 'Smelt'. A design dubbed 'Kingfish Upper' used a single wide inlet on the top of the forward fuselage, installing the cockpit within the inlet ramp. A configuration dubbed 'Herring' was similar to 'Smelt' but rotated the side ramp inlets roughly 90° into a less stealthy but aerodynamically more sensible configuration.

The ultimate Kingfish was Configuration 257, dating from July of 1959. It was essentially a halfway point between 'Smelt' and 'Herring' with 'cheek' inlets tipped up about 45°. The inlets were essentially rotated versions of the 2D external-internal compression inlets to be used on the North American F-108. To reduce the radar return from the inlet, the upper inlet side wall and the cowl lip back to the inlet throat were to be made from radar-transparent fibreglass. A metal grid of 1.5in by 1.5in would cover the inlet throat to block radar from passing down the inlet and reflecting off the compressor of the turbojets. This same solution would be used two decades later to radar-proof the inlets of the Lockheed F-117 'Nighthawk'.

And finally, at long last, the pilot was moved to the centreline. The cockpit would be pressurized and air conditioned to comfortable shirtsleeve conditions. In the event of an emergency, a slightly modified B-58 ejection capsule would carry the pilot to safety. Since the capsule was also pressurized, the pilot would, it was felt, not need to wear a pressure suit. The

Convair FISH Configuration 238

SCALE 1/165

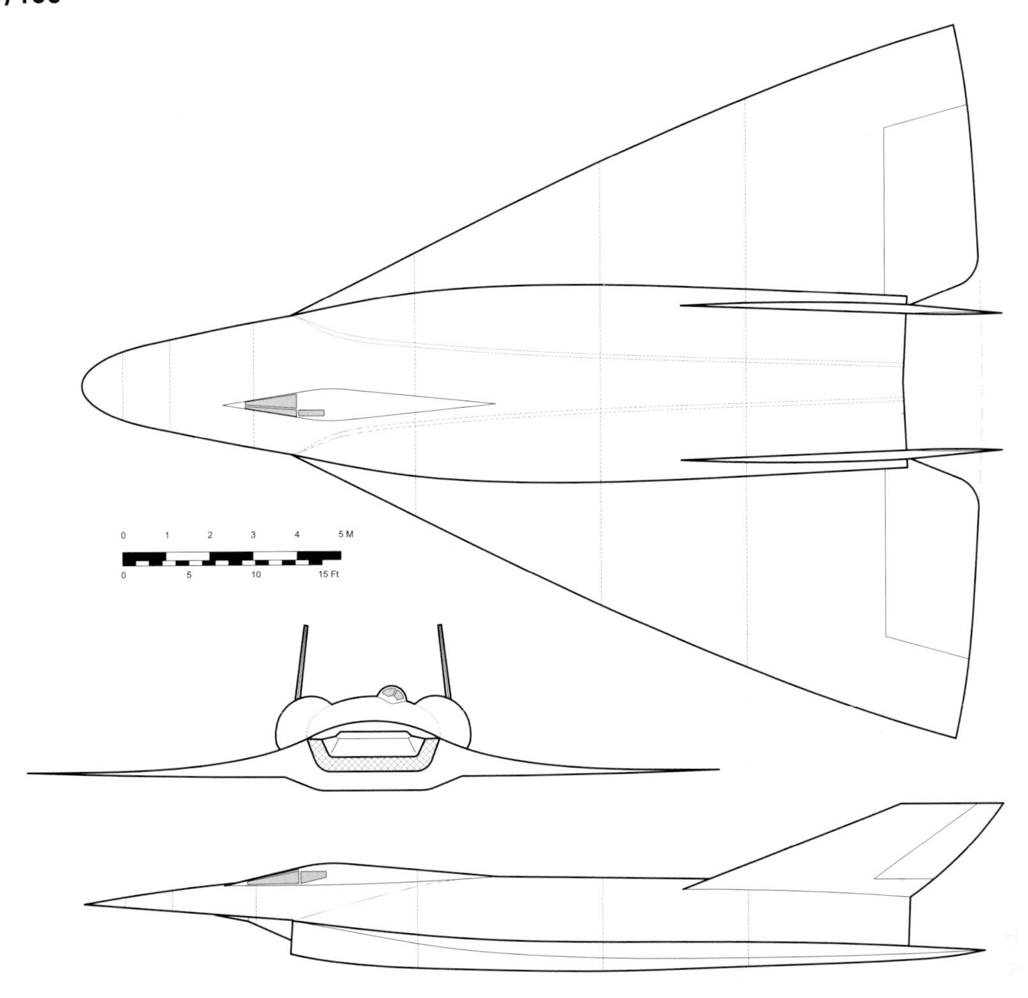

reconnaissance payload occupied the compartment directly behind the cockpit. Growth, either in crew or payload accommodation, seems unlikely given how compact and compressed everything was.

The wings were similar in planform to what had come before, with curved leading and trailing edges and pointed tips. The airfoil was a 2.5% modified bi-convex with sharp leading and trailing edges. Together with wing camber, low-speed aerodynamics would suffer, but cruise performance would be optimized. The vertical stabilizers were moved back out onto the wings at roughly the mid-span position. The stabilizers were fixed straight up and down… but since the wings had substantial dihedral; the wings and stabilizers did not form 90° corner reflectors. In any event the wings effectively shadowed the stabilizers from ground-based radar. The whole underside of the Kingfish was effectively one smooth, unbroken convex surface. The bulk of the effort of developing the Kingfish design was spent in radar studies of various large-scale models to find a practical configuration that would be essentially invisible to Soviet radar.

Skin honeycomb sandwich panels, bulkheads and spars were to be fabricated from PH15-7Mo stainless steel. Intermediate frames and bulkhead inner flanges, spar webs and fittings were to be made from RS-140 titanium. For particularly high temperature components of the airframe such as around the exhaust, the nickel-based Inconel 718 superalloy was planned. Pyroceram 9606 would stay the material of choice for the wing edge triangular inserts, held within a steel structure. In order to support thermal emissivity, the aircraft would be painted with something like Dow Corning XP5315 with 10% ceramic black. This paint would have an emissivity of more than 0.9 (meaning that it would be black in visible wavelengths) at above 600°F.

The Kingfish would be powered by two Pratt & Whitney JT-11 turbojets. This was the company designation for what would come to be officially designated the J58, the same engine used by the SR-71. Fuel would be conventional RJ-1. Provisions were made for in-flight refuelling from a KC-135A using a receptacle in the nose. The engines had rectangular 2D nozzles, separated by a partition. This was needed

Convair FISH Configuration 242-A
SCALE 1/165

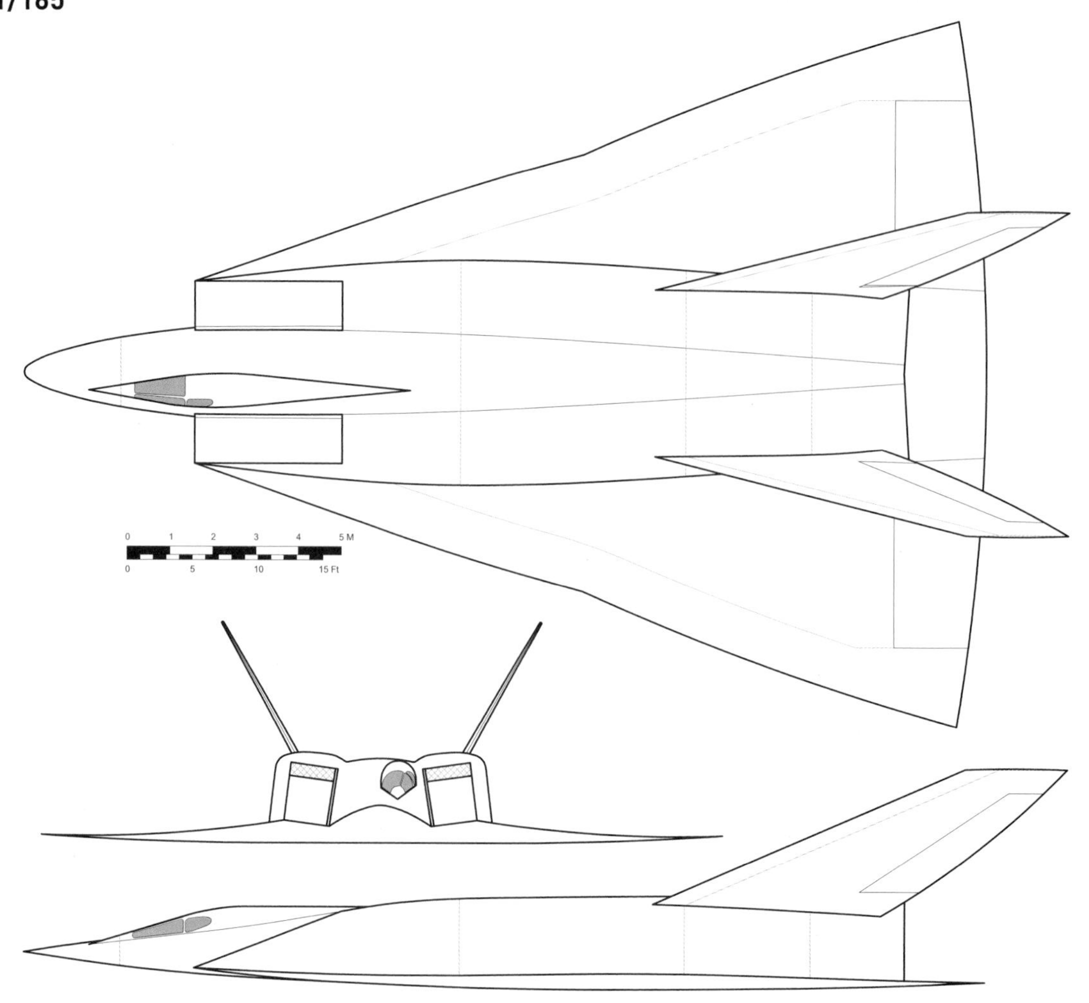

in the event of an engine-out; otherwise the exhaust from the functional engine would spread out over the wide rear of the fuselage, adding an unwanted yaw component to the thrust.

Wing loading would be relatively low at 56lb per square foot. Coupled with a decent thrust-to-weight ratio of 0.58, liftoff speed would be 138 knots, requiring a ground run of 1570ft, clearing a 50ft obstacle in 3530ft.

A standard mission would involve three legs and would require refuelling three times. Following takeoff from a base in the continental United States, the Kingfish would climb at 400 knots to around 76,000ft and accelerate to Mach 3.2. It would cruise at that speed for 3540 nautical miles (climbing to around 87,000ft), when it would decelerate to subsonic and descend to 35,000ft for refuelling. Then it would climb to 85,000ft and resume Mach 3.2 cruise as it overflew target territory, climbing to around 94,000ft and crossing 3430 nautical miles. During this 'tactical leg' of the mission the Kingfish would perform two 90° turns. Once again it would return to refuelling altitude and velocity and then climb to 74,000ft for a Mach 3.2 cruise home, climbing to around 87,000ft as it crossed 3645 nautical miles. Including distance crossed while climbing, descending and refuelling, total range of the aircraft would be 12,280 nautical miles, for a mission time of 8.17 hours.

While Convair Fort Worth staff were hard at work on Super Hustler, FISH and Kingfish, their compatriots at the San Diego facility were also busy – to a much lesser degree – on a series of high-altitude, high-speed reconnaissance planes of their own. While Fort Worth worked on designs for the USAF and the CIA, San Diego worked on designs for the Navy in a programme that has gone largely unreported. Project Hazel was a wide ranging but relatively small programme that ran from May 1958 into May 1959. In many respects some of the requirements paralleled those of the FISH studies, but Project Hazel included capabilities far beyond what FISH could attain. Curiously, there seems to have been no crossover between the efforts; the distance from San Diego to Fort Worth was too great.

General Dynamics Unstaged Special Purpose Super Hustler

SCALE 1/100

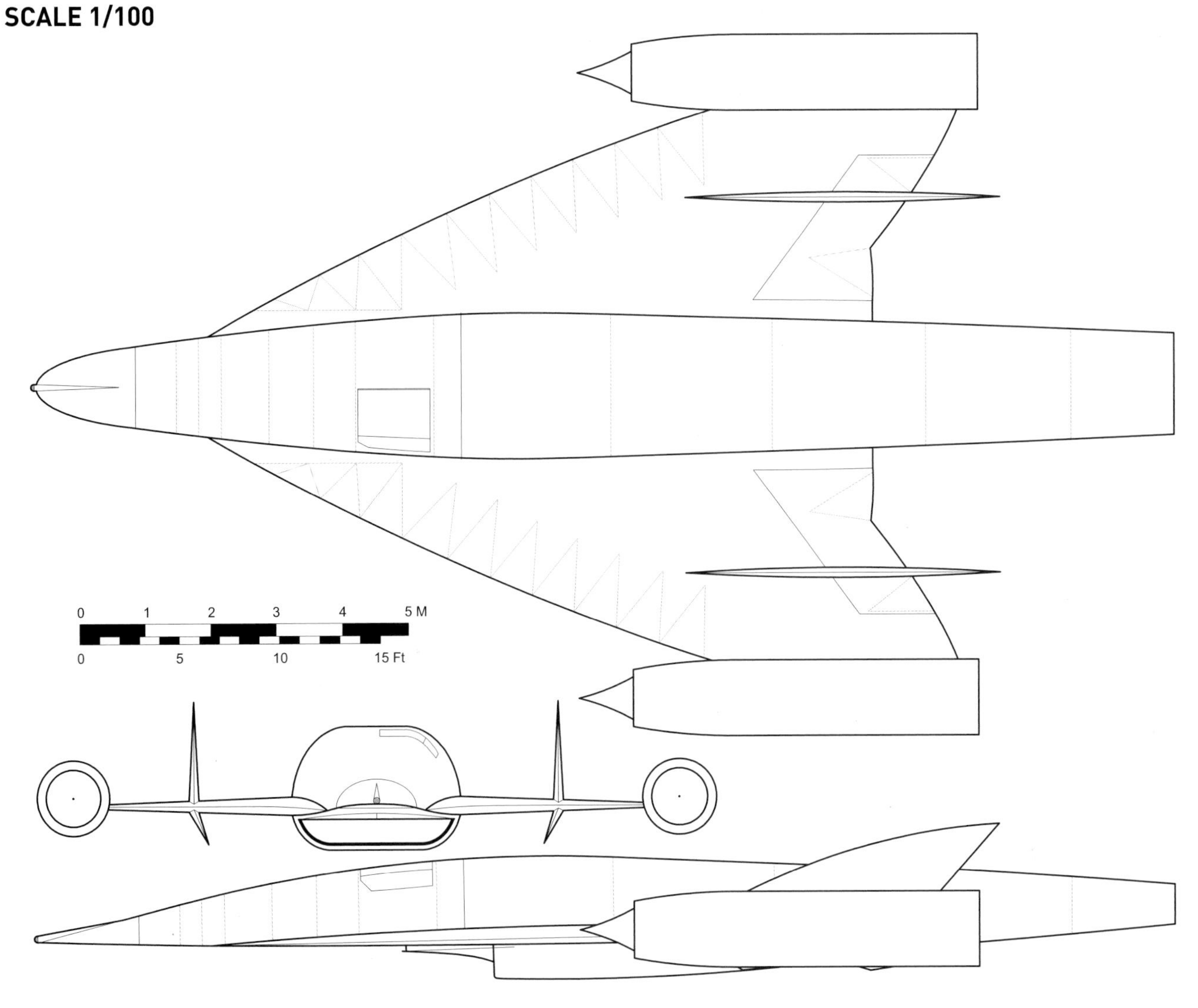

The main effort was reported in October 1958 and included a number of very different aircraft designs. Compared to the Super Hustler and FISH studies, the Hazel designs were fairly crude and do not seem to have been backed up with wind tunnel studies or detailed structural or aerodynamic analysis. The designs largely seemed to fall into two categories: first, highly swept delta wing designs somewhat similar to Convair's F-102, F-106 and B-58, built with conventional rigid metal structures; second, flying wings shaped like equilateral triangles made from 'inflatable' plastic structures.

The designs featured different numbers of engines, different engine locations and different propellants. JP-4 was considered along with liquid hydrogen and sportier high energy propellants such as pentaborane.

One of the more prominent Hazel designs was the MC-10, a Mach 3 ramjet-powered triangular flying wing aircraft. The MC-10 would be made of fibreglass and would base its rigidity not on a sturdy internal structure but on internal air pressure. In order to fly at the extreme altitudes proposed, wing loading would need to be kept low, meaning a minimal aircraft weight. To accomplish this, Convair used a concept that had been proven with the Goodyear 'Inflatoplanes'.

These Goodyear designs utilized a structural concept called 'airmat'. The outer skin of, say, a wing would be made of an impermeable, stout rubber-infused fabric. With just the skin, a wing would tend to inflate into the shape of a cylinder or a sphere, so the actual airfoil shape was maintained by a vast web of fibres within the wing that constrained the skin from inflating past the desired outer mold line. The concept worked extremely well; with substantially less air pressure than is required to fill a car tyre, a practical aircraft could be inflated to sufficient rigidity to fly. The airmat concept remains

General Dynamics KINGFISH Configuration 257

SCALE 1/120

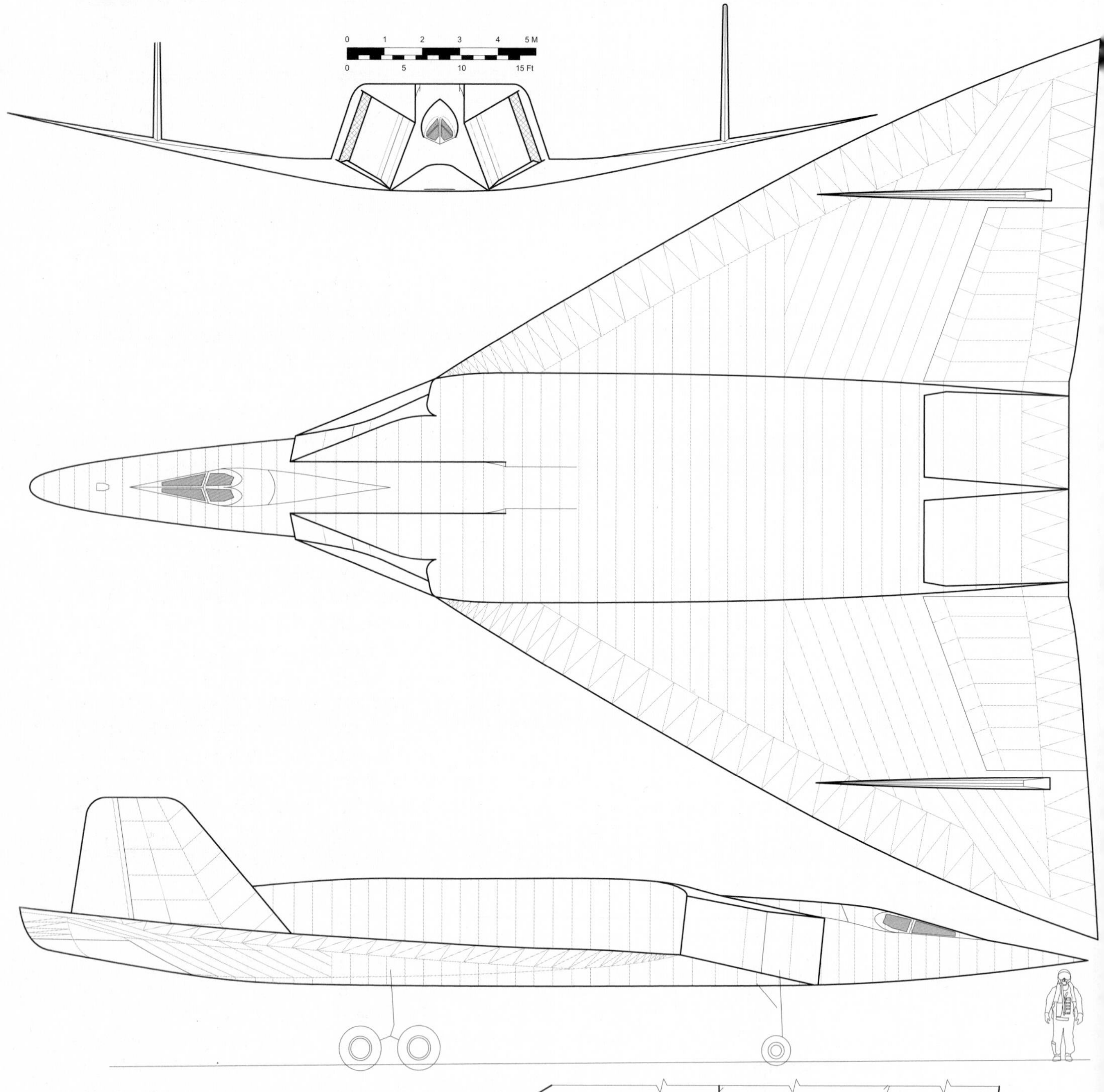

0 1 2 3 4 5 M
0 5 10 15 Ft

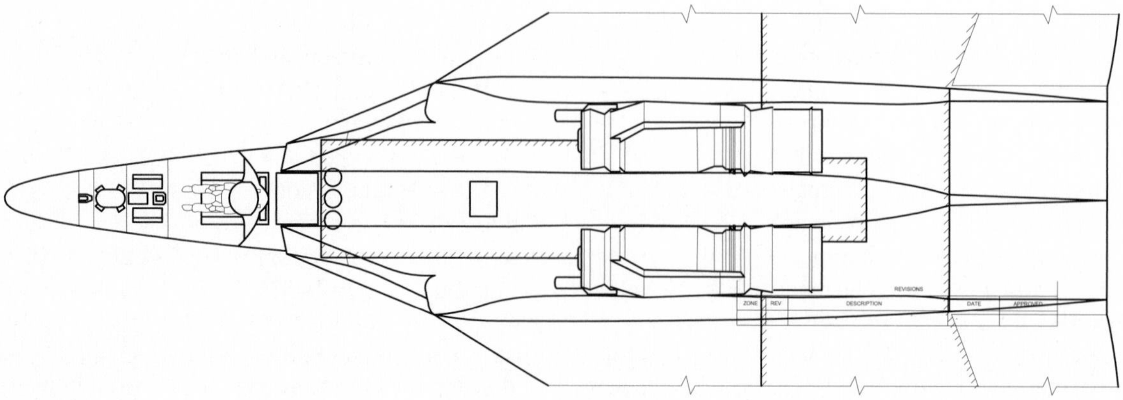

SCALE 1/144

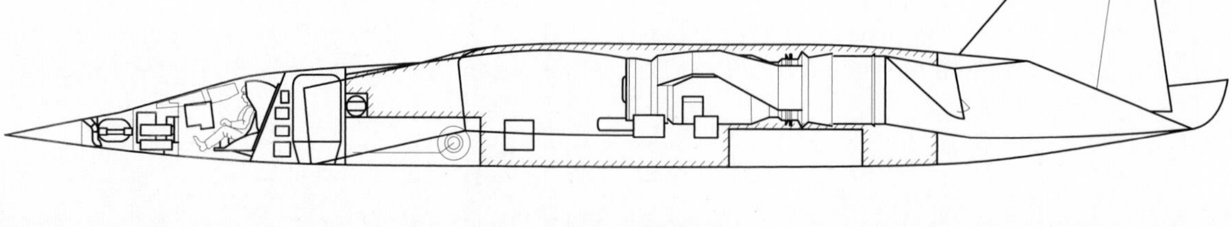

the strongest wing structure by weight. Goodyear's 'Inflatoplanes' were distinctly low-speed craft but the concept was valid for high-speed planes as well and gave the Convair designers a way to create a very lightweight high-altitude, high-speed design. That the fibreglass structure would not be as radar reflective as an equivalent metal structure... that was good too.

The MC-10 used fibreglass pressure-stabilized delta wings (silicone impregnated fibreglass fabric) containing a rigid pilot capsule, topped with a single remarkably large ramjet engine. The gigantic Marquardt ramjet engine was itself made largely out of a type of plastic honeycomb, a high temperature ceramic fibre impregnated with phenolic resin similar to the commercially available 'Refrasil'. The fuel system, flame holder, cooling shroud and engine mounts were to be made of metal.

The central section of the aircraft was a rigid structure made of laminated fibreglass. Aft of the pilot's capsule was a bay containing the recon equipment. Aft of that was the fuel tank, containing the pentaborane fuel. At the time (late 1950s) pentaborane was a popular subject of study as an aerospace fuel, both for jets and rockets, due to its high performance potential. This 'zip fuel' was originally the intended fuel of the B-70 bomber, but the B-70 eventually went with a more conventional hydrocarbon fuel, partially due to lower cost but mostly due to the fact that pentaborane is a somewhat insane fuel.

Performance is theoretically good – potentially game-changingly good due to a specific energy nearly twice that of kerosene – but it also bursts into green flame on contact with air, which makes the smallest leak or spill monumentally dangerous. It reacts spontaneously with water and halocarbon fire suppressants. Not only are the combustion products chemically toxic, the pentaborane itself is a toxin on the scale as some nerve gases. The safest way to put out a pentaborane fire is to watch it on the news from another state, preferably upwind. And from a performance standpoint, the combustion products include various boron compounds. Boron trioxide will not only condense out onto turbine blades and exhaust nozzles as a thick glass-like mess, it will suck thermal energy out of the surrounding exhaust gas as it does so. Boron carbides will also condense out of the exhaust in the form of small, sharp, incredibly hard and abrasive particles. None of this is good for the long term operation of any engine. And on top of that, the solid particulates form a clearly visible smoke trail pointing directly at the aircraft.

As a pure ramjet aircraft, the MC-10 was incapable of launching itself to high speed and altitude. Ramjet engines are capable of incredible performance including great range when they are flown at their optimum design point, but ramjets are very sensitive and small divergences from the optimum altitude and airspeed quickly degrade performance. So the MC-10 needed to be a 'staged' aircraft. Convair studied, and rejected, several approaches, firstly high-altitude balloons. Considered practical for a test programme, using vast helium balloons to haul the aircraft to high altitude prior to release was considered unworkable from an operational or tactical standpoint. Second was rocket launch. This would impart stresses on the aircraft that were too high to tolerate, as it would necessarily see high speed and acceleration while still in the lower atmosphere. Third, a specially designed high-speed airbreathing carrier aircraft was rejected by the Navy, probably for being simply too much of a development effort.

In the end, the baseline launch system was to carry the MC-10 to altitude atop a B-52. When properly positioned, the MC-10 would separate and be boosted higher and further yet by rocket boosters. Interestingly, it was not expected that the MC-10 would be mechanically latched to the B-52. Instead, the B-52 would be fitted with a launch platform with the same planform as the MC-10, and with a concave upper surface that closely matched the convex outer geometry of the MC-10s lower surface. The MC-10 would fit snugly into this and be held in place via suction, allowing the recon aircraft to be held securely without the need for penetrations into its inflatable skin, as well as providing a strong hold that could be released by simply turning off a switch. Most importantly, this system eliminated aerodynamic loads on the inflatable wing structure during the turbulent, bumpy low-and-slow portion of the initial flight. The upper surface of the support platform would be made of a honeycomb sandwich with a perforated outer skin to get a vacuum grip on the lower surface of the MC-10's wing.

Just before separation, the B-52 would slow to minimum safe airspeed, which at 45,000ft was 200 knots. A rocket booster system attached to the underside of the MC-10 – projecting through the launch platform via a cutout in the structure – would boost the aircraft to 125,000ft altitude and around 3150ft per second (more than 2100mph). At that point the ramjet-powered cruise would begin.

It should be noted that the B-52 is, of course, a US Air Force aircraft, not US Navy. Left unclear was whether Convair suggested that the Navy rely on co-operation with the Air Force for launches, or if the Navy would buy its own B-52s. In that case, it's interesting to ponder how those Navy B-52 carrier aircraft might have been painted, and whether the Navy might have procured more B-52s to serve other roles. Prior to B-52 baselining Convair considered using the B-36, which was quite obsolete by that point.

Part of the Navy Project Hazel study included the notion of launching and recovering these high-

Convair 'HAZEL' MC-10

SCALE 1/155

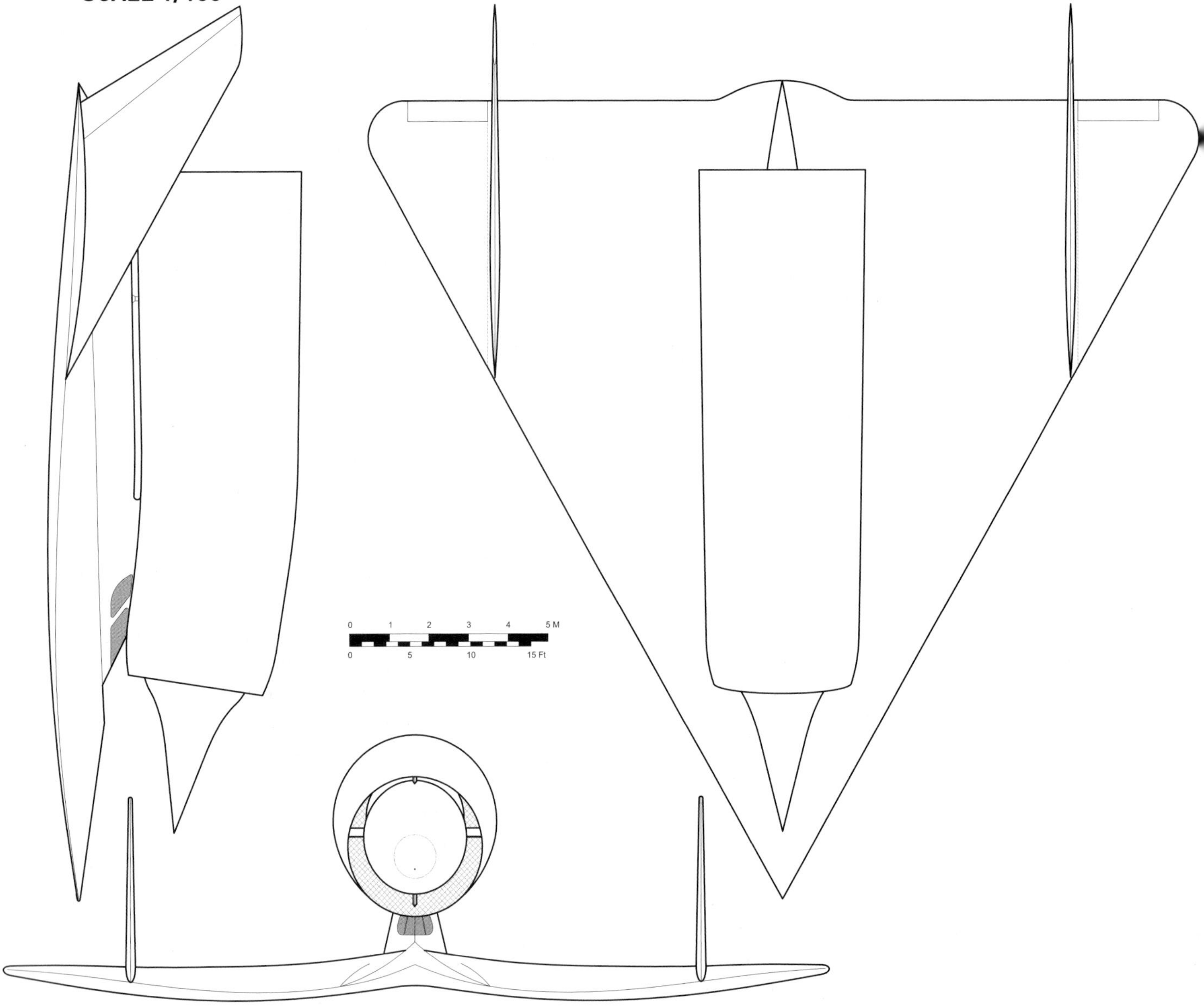

altitude Mach 3 aircraft from submarines. While the MC-10 would not have worked for that application (the ramjet engine was simply too big, and few enough sailors would want to spend weeks at a time trapped underwater in a metal tube knowing that there were tons of pentaborane fuel mere feet away just itching for the chance to spring a tiny leak and fill the sub with green dragon fire), the inflatable fabric structure was considered promising for other designs. The MC-10 could, at least theoretically, be deflated and partially 'rolled up' for storage after a flight. Only the ramjet engine and the central cockpit/equipment bay/fuel tank structures would have been rigid.

At the end of the mission, the MC-10 would glide to a landing, alighting upon skids. The skids would allow for landings on runways, snow or water, handy given that this was a Navy aircraft. So long as the aircraft was intact, it would of course float. In the event of a damaged aircraft, the pilot could eject through the floor of his capsule. A hatch would simply hinge at the front, dropping the pilot's seat down and back as through a trap door. Convair studied alternate schemes, including one that would have undoubtedly have made for some of the most unnerving piloting in history: the pilot would wear a 'frog-man' suit, would pilot in a prone position... and would float suspended within water. The cockpit would be a close-fitting shell completely enclosing him and filled with water. This would allow the pilot to withstand, it was believed, up to 80 g's of acceleration. Consequently, no ejection system was required because it was believed that under most circumstances, the low mass and high surface area of the fabric aircraft would allow it to plummet from the sky slowly enough for the floating pilot to survive any crash.

Another approach was embodied in the M-124, M-125 and M-126 series. These designs were more conventional in appearance than the MC-10, using delta wings, a distinct fuselage and a vertical tail similar in layout to other Convair delta winged aircraft of the time. They had two modestly sized ramjets instead of one giant engine located over the fuselage. Most of these designs had small wings with ramjets located on the wingtips, but there were some variations in engine location. The design studies were conducted with certain performance ground rules... a range of 4000 nautical miles, a cruising speed of Mach 3 at an altitude of 100,000ft, a crew of one, 500lb of payload and a conventional metal structure.

The M-124A was the first design, but it is not clear if it was the preferred design. The M-124A was similar in configuration to, but smaller than, the Convair F-106A. It was in fact surprisingly small, being barely large enough for the pilot. At first glance the performance of the aircraft seems rather optimistic. It used pentaborane fuel, with all the associated cost and risk. The performance of the vehicle was quite good (including the highest cruising altitude), but that performance benefit was stated as insufficient to offset the added headache of the pentaborane.

In order to get the M-124A up to speed, it would need to be boosted. Launch was to be either from the ground or from under the wing of a B-52. Ground launch involved the use of four solid rocket motors. Included were a large first stage booster taken from a Minuteman ICBM attached to the underside of the rear fuselage; a smaller solid rocket on the underside centerline as a second stage, and two smaller still boosters flanking the second stage to serve as a third stage.

This vehicle would launch vertically like the space Shuttle and was not recommended. Another approach involved using liquid rockets to achieve the same end. Three boosters were used, each including a single Bell Model 8048 rocket engine (aka the XLR81, used on the Agena upper stage from 1959 to 1961) and tanks containing the nitric acid oxidizer and unsymmetrical dimethylhydrazine fuel. All three rockets would ignite for a vertical launch, then the two smaller boosters would drop away, leaving the larger central booster to continue to burn until the M-124A reached the target altitude and speed.

The preferred launch option was using a B-52 to carry either one or two M-124s at a time. The B-52 could greatly extend the range of the aircraft over ground-launched versions. The M-124 would be carried under a wing pylon in much the same manner as the Hound Dog missile. After release, it would use a trio of solid rocket motors (the second and third stage rockets from the solid rocket VTO version) to accelerate to ramjet speed. This approach is strikingly similar to what would follow in actual practice some years later when Lockheed used B-52s to launch D-21B drones. The M-124A was just small enough so that no wingtip, tail surface or nose folding would be required to clear the ground or B-52 components.

It was proposed that for training purposes a version of the M-124A would be built where the wingtip ramjets would be replaced with J85 turbojets burning conventional JP-4. Training flights would involve only the lower end of aircraft performance, topping out at Mach 0.7 and around 50,000ft altitude. Range for the training vehicle would be only 580 nautical miles.

A proposed step in the development of the M-124A was the M-124B. This would have used the same airframe but different ramjet engines, designed to use conventional JP-4 fuel rather than pentaborane. The only visual difference was that the engines were slightly longer to accommodate the slightly slower and less energetic combustion of the JP-4. The fuel tanks in the fuselage would have remained the same. Performance would have suffered: cruise altitude would have been 10,000ft lower and range would have been cut by 1000 miles. But this vehicle would have allowed the design and the airframe to be flown with a far less problematic – and far cheaper – fuel.

Another related design was the P-124C. This was the same basic configuration modified for SF-1 (liquid hydrogen) fuel; unsurprisingly, this resulted in the fuselage being far fatter than either the M-124 or M-124B as well as stretched somewhat. The hydrogen burning ramjets are quite short in length compared to the slower-combusting alternate fuels. Once again it was found to be not worth the extra trouble that using liquid hydrogen would bring.

It does make an interesting comparison with the similarly configured Lockheed CL-400. With less than half the length, around 13% the gross weight (not counting the weight of carrier aircraft and acceleration rockets) and with about 14% the fuel load, it was projected to fly higher, further and faster. Whether it would have retained those performance characteristics once detailed design work began is doubtful.

The M-125 was a little different. It was designed from the outset to have minimum radar cross section, with a larger wing and the ramjets relocated to roughly mid span above each wing. This would shield the ramjets from ground-based radar systems looking upwards. However, analysis (including model testing) suggested that the reduction in radar cross section from the baseline configuration would not be enough to make up for a reduction in altitude performance.

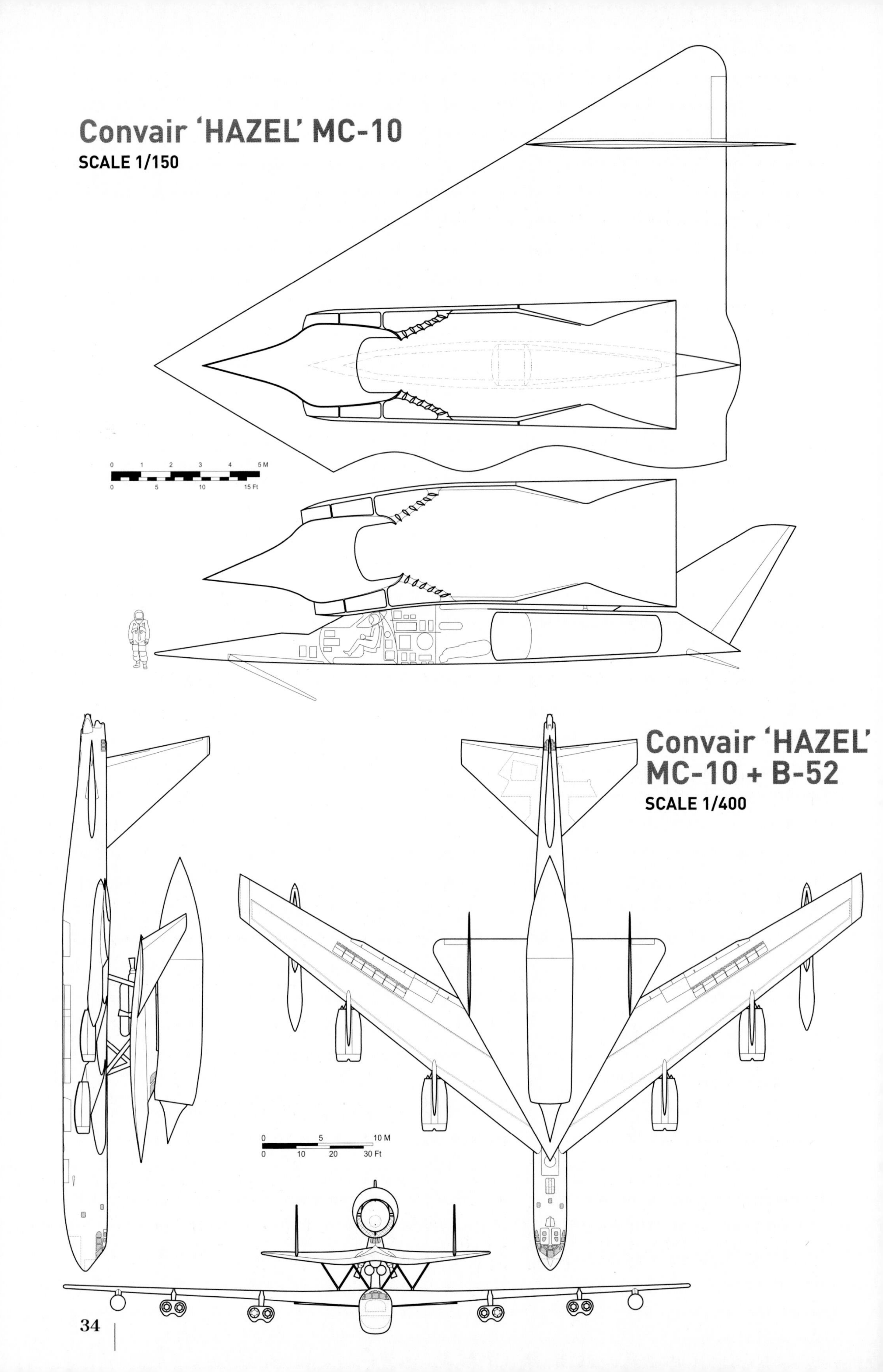
Convair 'HAZEL' MC-10
SCALE 1/150
0 1 2 3 4 5 M
0 5 10 15 Ft
Convair 'HAZEL' MC-10 + B-52
SCALE 1/400
0 5 10 M
0 10 20 30 Ft

Convair M-124A

SCALE 1/100

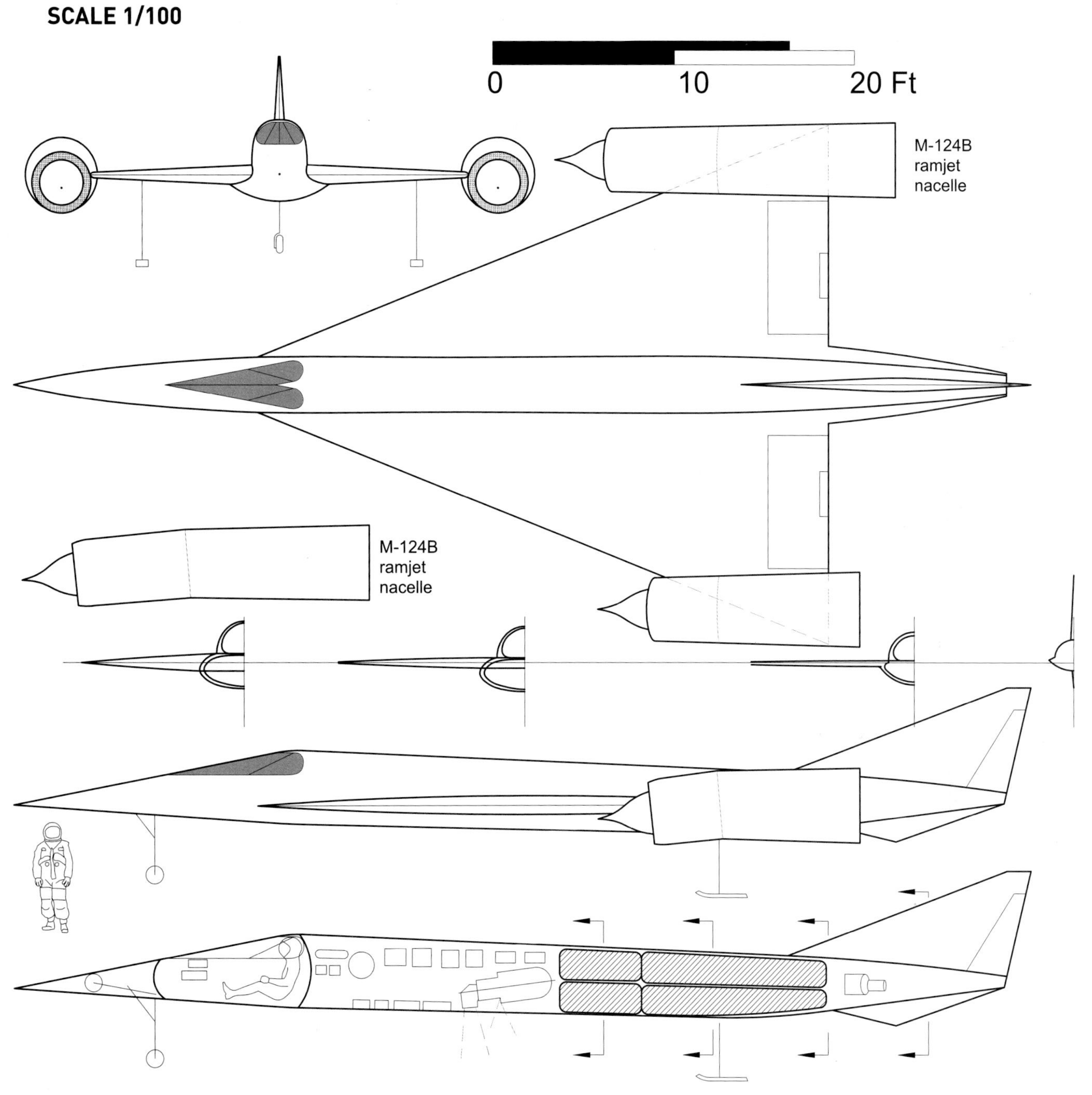

About half of the fuselage was taken up with a two-compartment fuel tank. The fuel, as with the M-124A, was pentaborane.

The M-126 was a high performance JP-4 version. It was similar in configuration to the M-124B, but rather than being simply a JP-4 fuelled version of an aircraft optimized for pentaborane, the M-126 was optimized for JP-4. As a consequence the fuselage, to be made largely from titanium, was fatter and longer for greater fuel volume. The configuration was much like that of the M-124B but the camera payload was relocated to the rear fuselage. The wings were also substantially reduced in size, now appearing almost as afterthoughts.

Since this study was being made for the Navy, launch from naval vessels was examined for the M-126. It was small enough that multiple vehicles could be carried by and launched from both modified freighters and specially made submarines. In both cases the aircraft would be launched vertically using either the liquid or solid rocket boosters proposed for the M-124. In the case of submarine launch, the aircraft would need to be modified to incorporate folding wings in order to fit within storage tubes some 13.35ft in diameter. The simple diagram showed a rectangular cross-section submarine carrying the storage tubes at an angle to fit within the hull, four tubes side by side, and at least three

Convair M-124A + B-52
SCALE 1/200

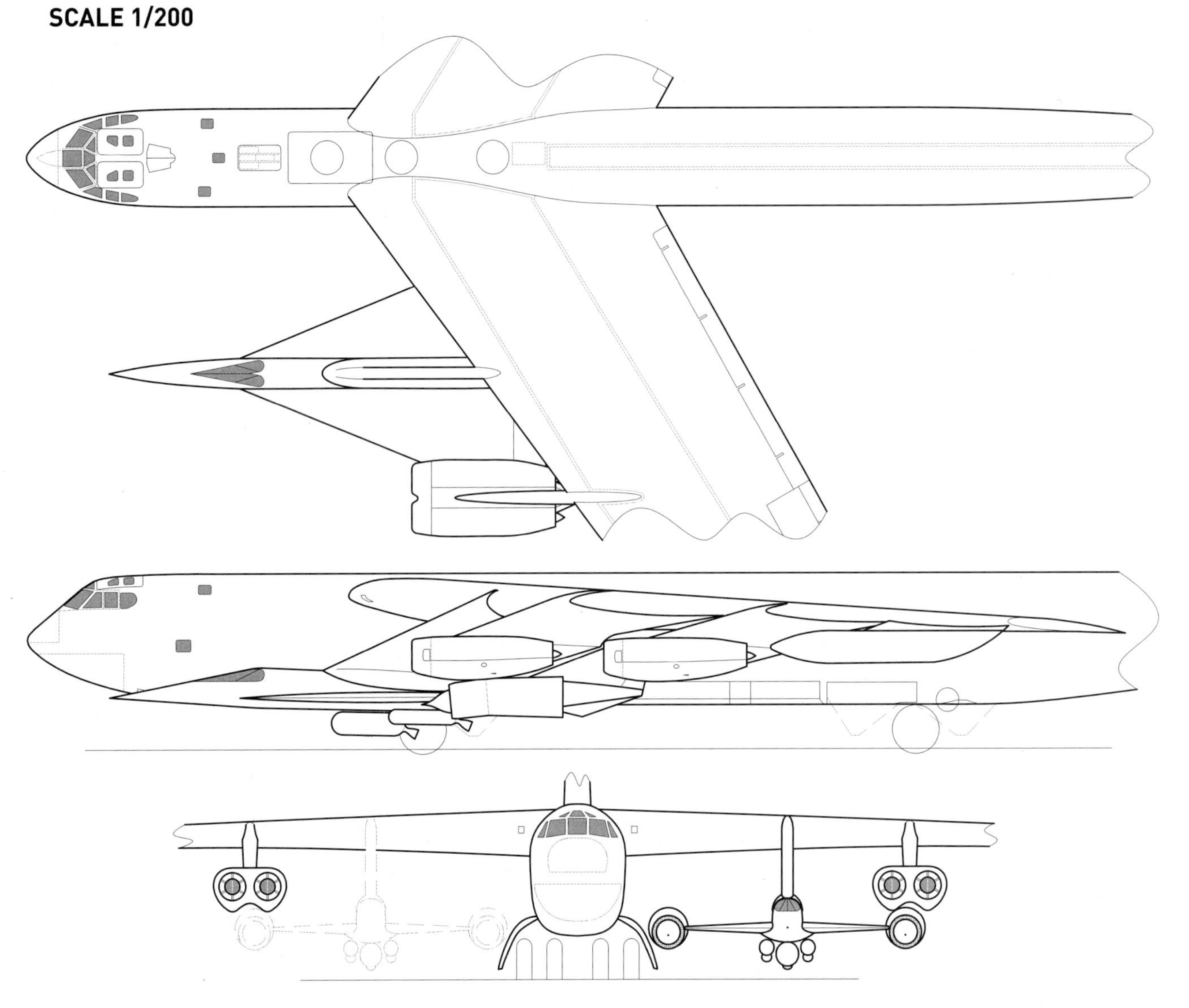

rows of tubes, with room within the hull for five tubes for a total of 20 aircraft. There would be no provision on either the surface vessel or the submarine for recovery of the aircraft; they would need to land either on an aircraft carrier or at an air base.

Convair's recon B-58s

While Convair lost out to Lockheed, the company did not give up on the idea of supersonic reconnaissance aircraft. The idea recurred several times in several ways. In the early 1960s, the idea arose of using the B-58 in a whole new way to spy on the enemy.

In November of 1960 Convair proposed to develop a supersonic transport variant of the B-58. During the early Sixties there was serious interest in supersonic passenger travel as the natural extension of the Jet Age, and while most of the major aircraft companies proposed to dive straight in to building full-sized passenger transports capable of Mach 2 or better, Convair had the idea of building a relatively small SST based on the B-58.

It was understood that the Model 62 (AKA 58-9) SST would not be a commercially competitive – or even a commercially viable – passenger transport… but it would be a good testbed for passenger reactions to supersonic travel. It was essentially a stretched version of the proposed B-58C, a tactical bomber/ interceptor that used four non-afterburning Pratt & Whitney JT11-B2 (J58) engines in a somewhat different configuration to the B-58A.

A dozen Model 62s would fly from airport to airport shedding sonic booms across the land, showing either that such noises were well tolerated by the public or that they weren't. As valid as the idea of an SST demonstrator was, aerodynamics, materials and engine technologies were rapidly progressing far beyond the relevance of a subscale aircraft and interest in the Model 62 quickly faded.

Convair P-124C

SCALE 1/144

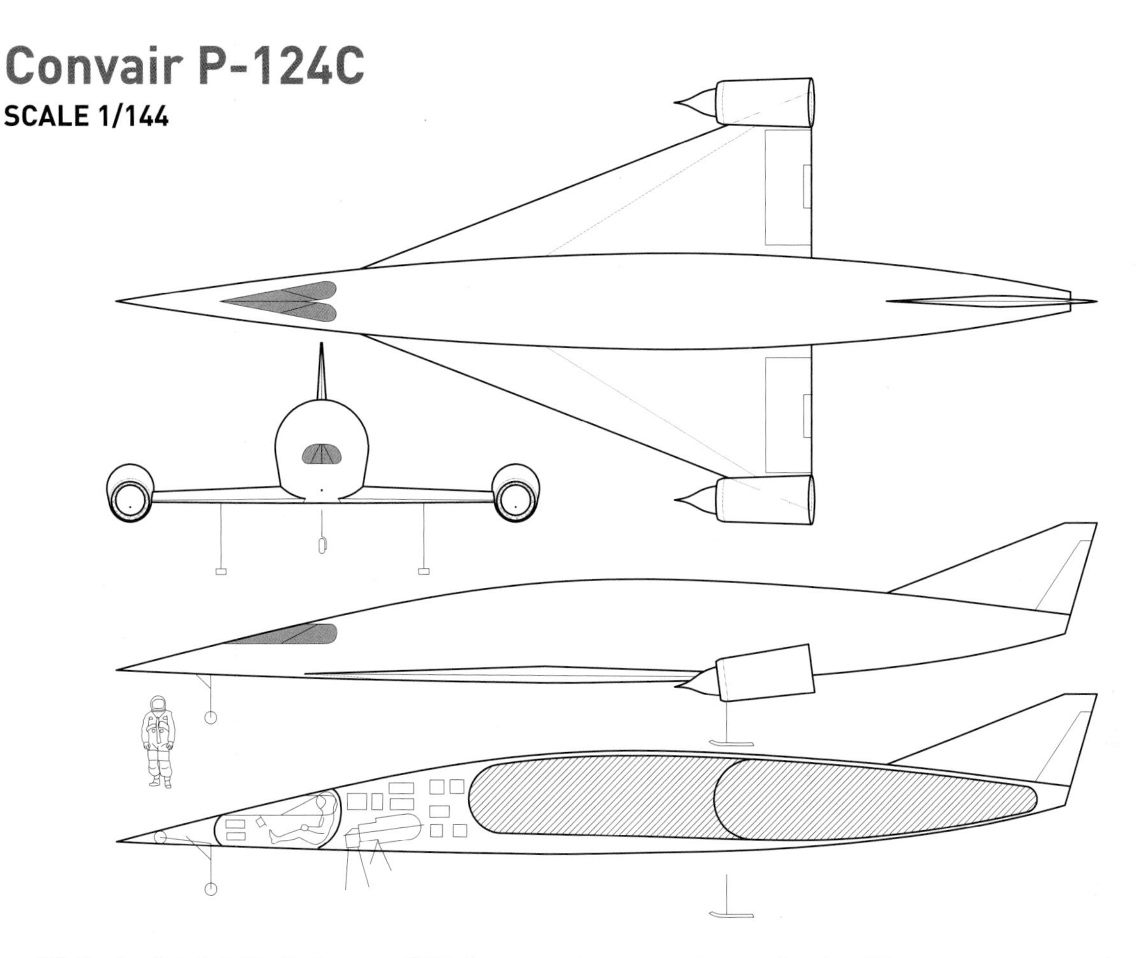

While the Model 62 died as an SST demonstrator, the basic idea of a B-58 derived transport for military purposes hung on for several more years. In September of 1961, Convair – now General Dynamics – released a design for a Special Purpose Aircraft, essentially the product of mating the Model 62 SST demonstrator with a reconnaissance equipment supply store. Passenger transport became a lesser concern, and instead the fuselage was packed with sensors and eavesdropping equipment. Consequently the fuselage shrank in length substantially, but grew in width. The Model RC/80, as it was called, could be used as a recon plane or a carrier of high-value cargo or a limited number of passengers.

Recon was apparently the primary role; while not as fast or as high flying as the SR-71, the RC/80 would be able to carry substantially greater payloads, as well as a larger crew to man multiple stations. The RC/80 would fly along the borders of areas of interest and peer sideways using the 'passenger windows' on the starboard side of the fuselage. The RC/80 would not overfly Soviet territory because it was neither stealthy nor did it fly beyond the reach of interceptors or missiles. But it could carry a relatively gigantic payload of 11 stations of optical, IR and UV sensors as well as side-looking radar. And while it flew its mission it would bring along not just sensors but a substantial crew: a pilot, co-pilot and navigator to fly the aircraft, along with a side looking radar operator, an electronics intelligence operator and two signal monitors (tactical and coordinator).

This crew would, unlike the crews of the U-2, A-12 and SR-71, be able to get up, move around and visit a galley and a restroom. A normal mission would involve a Mach 2.4 cruise on a leg of 2130 nautical miles followed by inflight refuelling, a further Mach 2.4 cruise to the destination, some 3840 nautical miles from the launch point. There the RC/80 would loiter for an hour at Mach 0.92, then return home with one further refuelling.

The passenger version of the RC/80 would look externally indistinguishable from the recon version apart from the side looking radar, and would have been able to carry 30 passengers. This is far too small a number to be commercially viable, but may have been useful for sending experts, officials and politicians far away relatively quickly. The cargo variant of the RC/80 was slightly different, mostly in terms of having a slightly longer fuselage with cargo loading doors on the port side of the forward and rear fuselage. But the Special Purpose Aircraft would at least look externally like a transport aircraft, perhaps fooling the Soviets for all of a few minutes about what its true purpose was. The RC/80 would have fulfilled a similar role to that of the RC-135, a series of reconnaissance versions of the venerable KC-135 tanker/cargo airlifter derived from the Boeing 707, and still in use today.

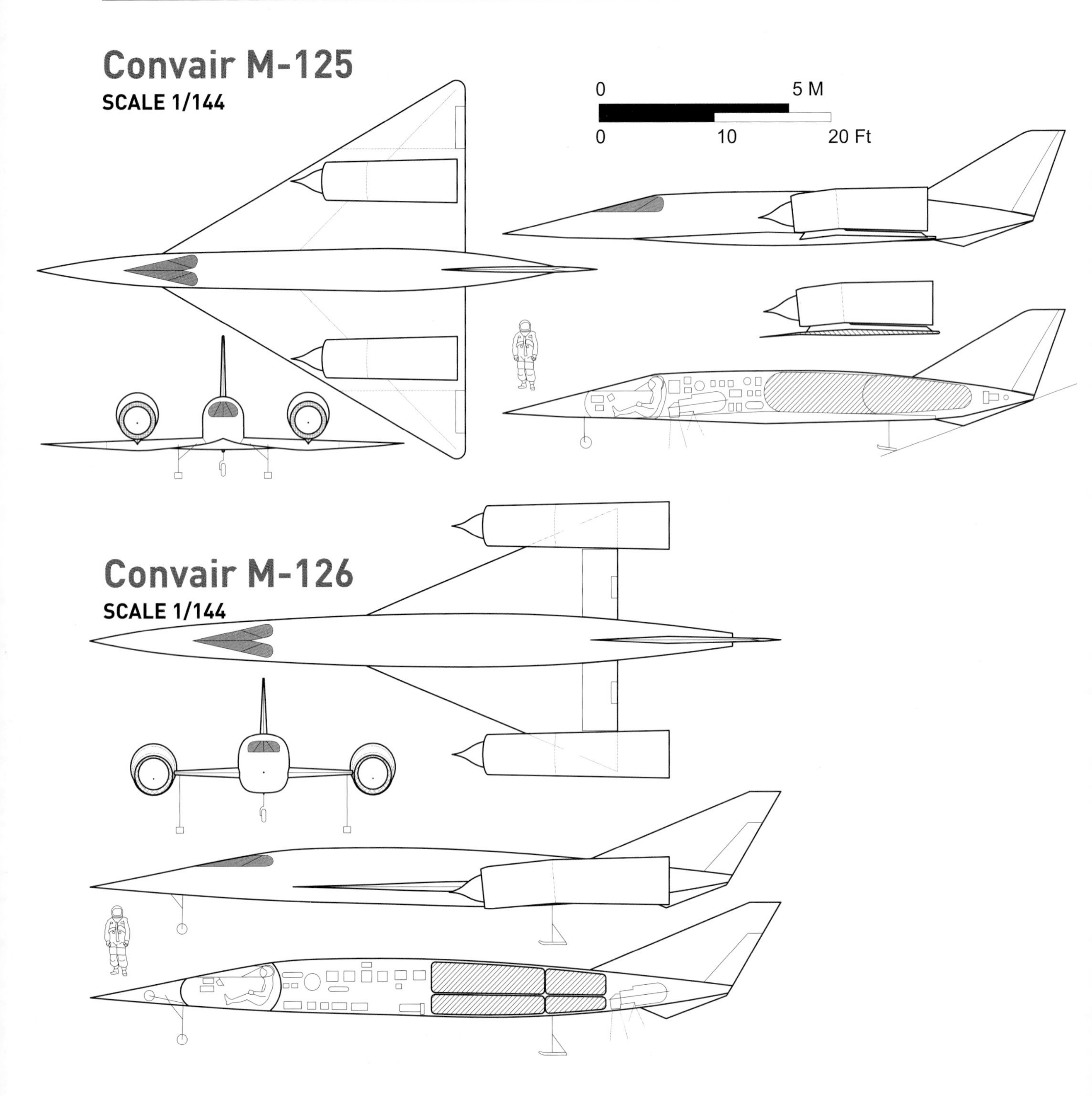

By May 1962, the QRC-182 emerged as a refined version of the RC/80. The fuselage was lengthened somewhat (about 10ft) and the cockpit canopy was greatly changed (the RC/80 canopy resembled that of the B-70, while the QRC-182 resembled that of the B-1). The fuel and gross weights were increased notably. The QRC-182-2 could still carry an optional underslung pod, but this pod carried additional cargo, not fuel. The added drag and weight would of course reduce range and altitude, though cruise speed was the same at Mach 2.4.

Passenger transport seems to have been a lesser requirement for the QRC-182, with available drawings showing either the recon 'special purpose' version or a cargo version. Sizable cargoes were not possible, as the internal volume was divided; the central fuselage fuel tank over the wing was split down the middle and equipped with a walkway to permit access from the forward cargo bay to the aft cargo bay. Additionally, the cargo doors (one fore, one aft) were on the side of the fuselage, rather than nose or tail.

For reconnaissance missions, a payload layout and crew complement just like that of the RC/80 was proposed. Available performance charts indicate that the QRC-182-2 was, unlike the RC/80, intended to

Convair QRC-182-2

SCALE 1/225

Convair RC/80

SCALE 1/225

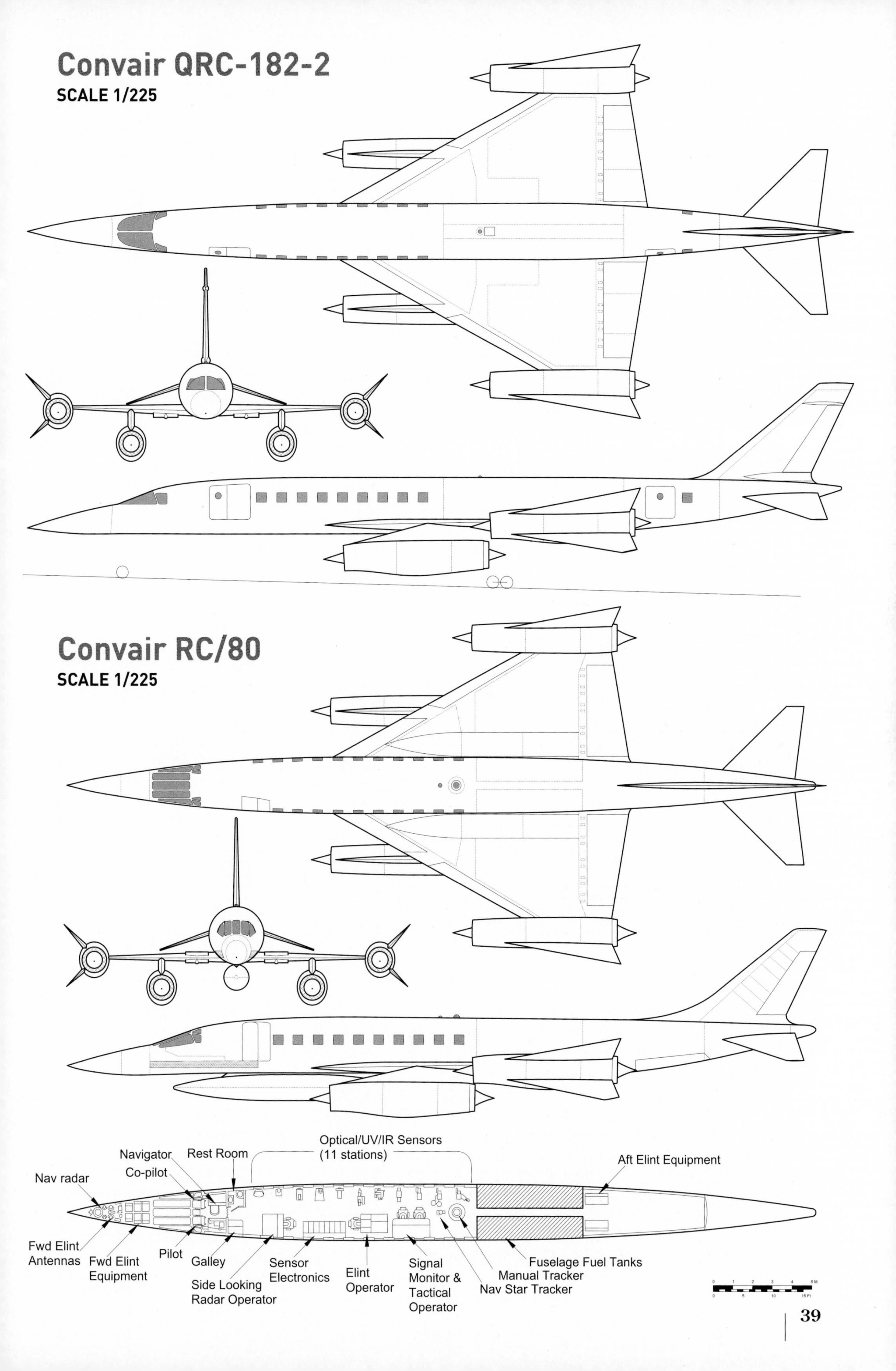

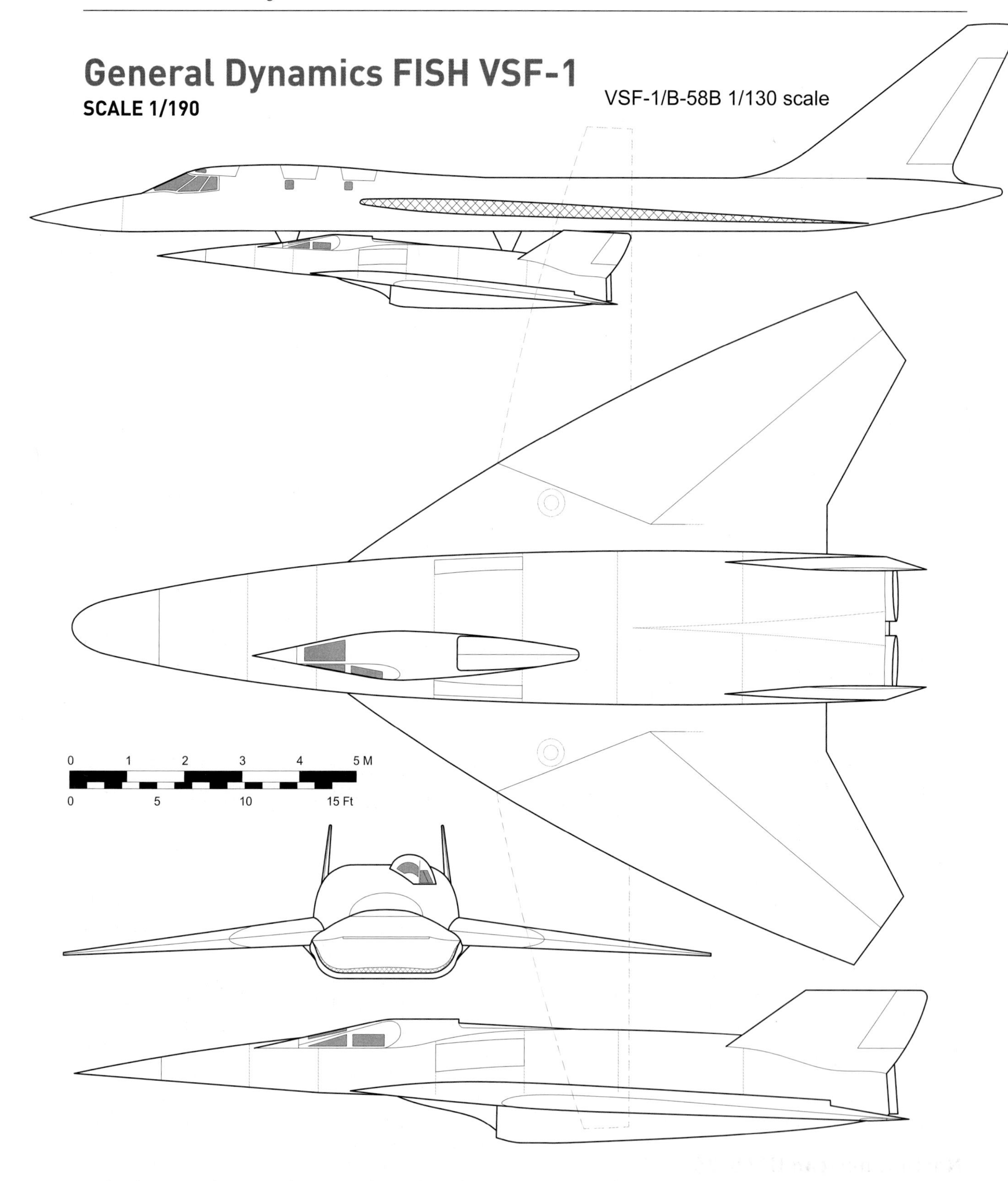

maintain a Mach 2.4 cruise throughout the mission except during refuelling. It does not seem that it was intended to loiter at subsonic speeds, but to cover great distance at speed. Refuelling in flight twice, it was expected to travel 8250 nautical miles in 6.4 hours, carrying 10,960lb of payload.

Several configurations were studied, including modifications to the planform to include leading and trailing edge root extensions, and a version where the tailplanes were replaced by canards.

Variable sweep FISH

The development and subsequent demise of Kingfish did not mean the permanent death of FISH, as the concept still had proponents at Convair well after the A-12 was selected. In November of 1963, a

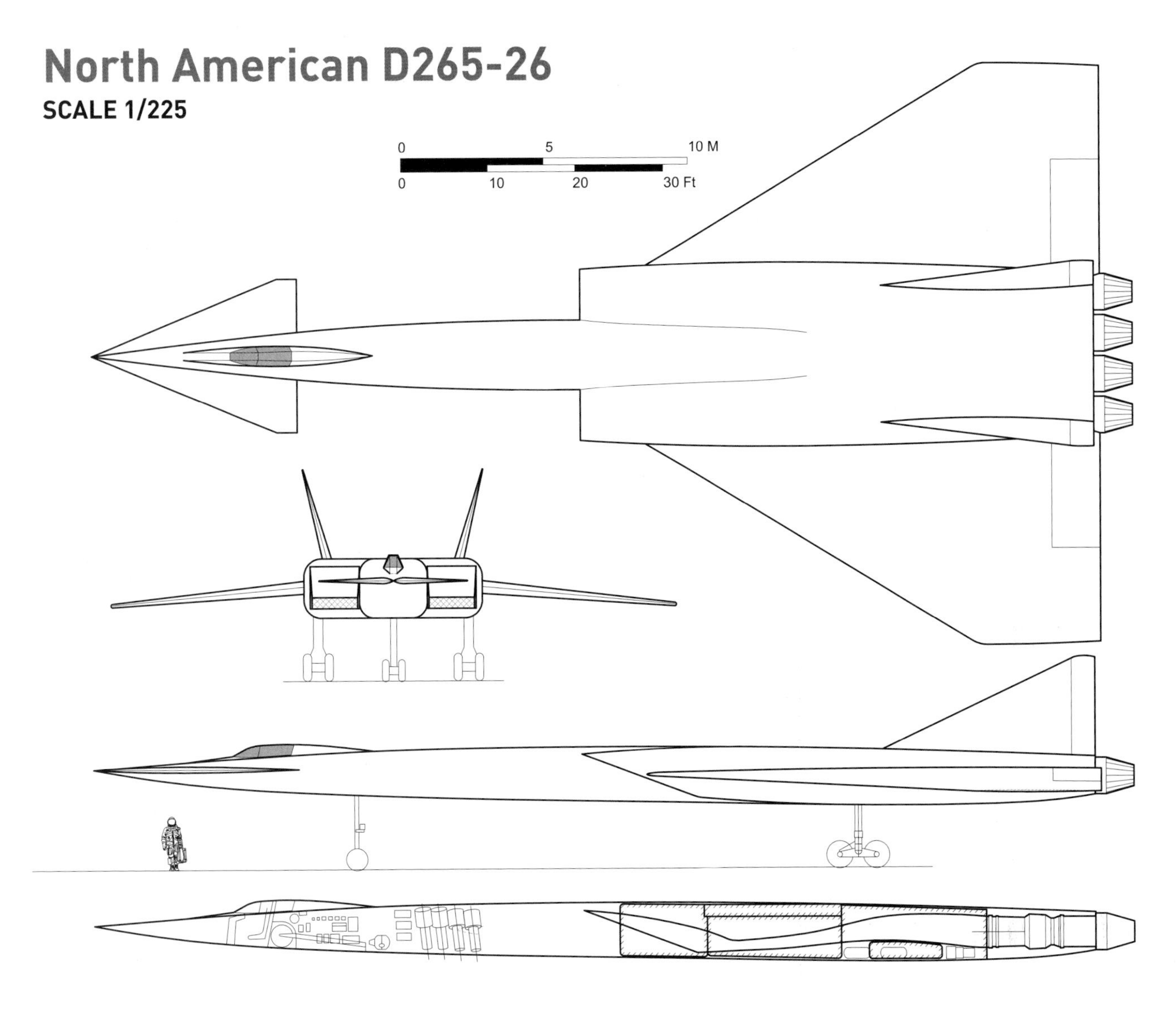

brief study was made of the VSF-1, a version of the Configuration 234 FISH with variable sweep wings. Variable sweep was the flavour of the moment in the USAF and the US Navy, with a wide range of aircraft being proposed with that feature (the F-111 being the one design of the time to actually see service). The VSF-1 was meant not to compete with but to replace the A-12 and it was thought that, somehow, giving the FISH improved low-speed flight characteristics would make the concept appealing to the powers that be. However, work did not progress very far. Replacing the A-12 was not to be an easy task.

North American D265-26

As described previously, in 1955 the Air Force asked for submissions to Weapons System 118P, a manned reconnaissance aircraft. North American Aviation submitted a design for WS 118P Phase II ½ that would somewhat replicate the capabilities that Convair and Lockheed were working on at the time. The D265-26 design was a four-turbojet manned aircraft with a configuration much like a combination of the contemporary North American F-108 and the later North American B-70 bomber… canards, clipped delta wings and twin tails. The all-moving canards were mounted at the extreme nose, directly ahead of the side-mounted 2D fixed ramp inlets. The forward fuselage was given an oddly square cross section.

The engines would be four JP-4 burning General Electric X278 turbojets. These were theoretical engines, J79-X275A engines scaled up to 135%, with a consequent increase in maximum thrust to 28,540lb with afterburner. The aircraft was equipped with four main fuel tanks… a forward insulated fuselage sump tank with a capacity of 10,140 gallons, an aft insulated fuselage tank of 6560 gallons and two wing tanks of 2350 gallons each. The aircraft was equipped for inflight refuelling.

It would be made largely from AM350 stainless steel, with three spars for the wing inboard sections and brazed honeycomb outer wing panels. Honeycomb panels formed the primary elements of the canard and vertical stabilizers. The fuselage was to be a semi-monocoque structure again largely made of steel, with

North American Manned SM-64

SCALE 1/175

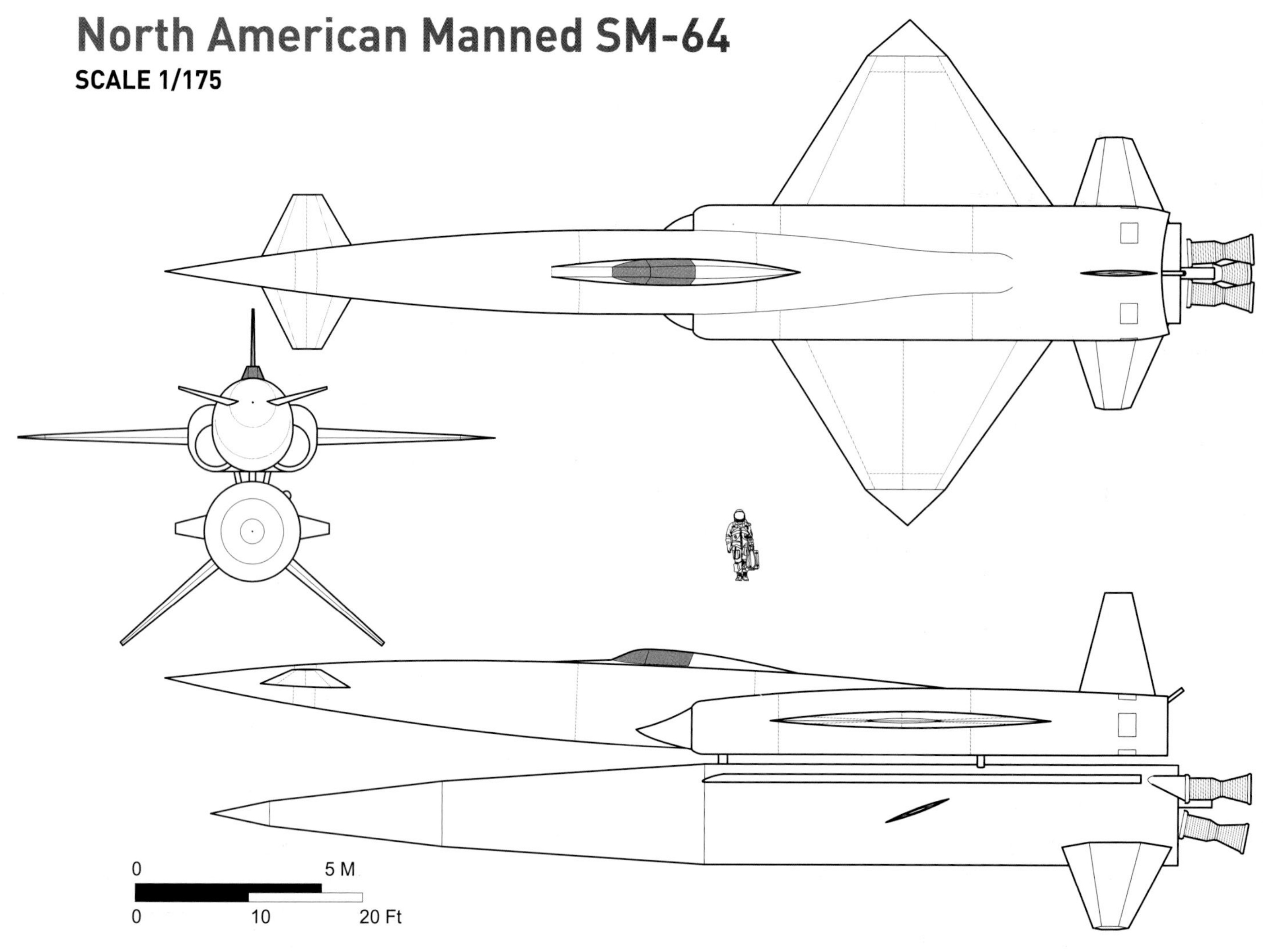

honeycomb construction for the inlets and ducting, and titanium skins on the forward fuselage.

North American estimated that the earliest date for an operational aircraft would be 1961.

Manned SM-64A

As part of the WS 118P study, North American were asked to look at the idea of modifying their SM-64A 'Navaho' cruise missile into a manned reconnaissance vehicle. Some data was produced, but so far no diagrams have come to light. It may be that the level of effort was little more than back-of-the-envelope, but it does make for an interesting – and perhaps obvious – idea.

The outer mold line of the Navaho would remain unchanged except for the addition of a canopy for the pilot. The warhead would of course be deleted, replaced with a cockpit and reconnaissance equipment; landing gear would be included. Presumably the landing gear would be the same as that planned for the XSM-64... a single pair of wheels for the nose gear, a skid under the rear fuselage and skids under the wingtips. The two XRJ47-W-7 ramjet engines and ducting would remain unchanged. A turbojet for landing would not be added; like the Space Shuttle, the vehicle would be expected to come in for a gliding landing. The structure would be changed to match normal piloted aircraft standards.

The recon vehicle would be boosted to 53,350ft and Mach 3.25 by a standard Navaho vertically launched rocket booster. The aircraft would then cruise for a range of 3830 nautical miles. The performance of this concept was not spectacular, especially in terms of altitude. Coupled with the cost and complexity associated with the vertical launch facilities and the expendable booster, it's unsurprising that the concept did not seem to merit a great deal of further examination. The diagram included here is provisional, using a North American D265-26 canopy added to the SM-64 over the missile's instrumentation bay. This would have left the fuel tanks unchanged, although the pilot's view would have not been terribly good.

3 Archangel

While Convair was working on the Super Hustler, Kelly Johnson of the Lockheed Skunk Works had his own ideas on high-altitude, high-speed reconnaissance platforms. After studying hydrogen fuel with the CL-325 and CL-400, he counselled against proceeding with it due to the disappointing range possible with a reasonably sized vehicle. But the end point of the CL-400 studies – the conventionally fueled CL-400-15JP – indicated that the sort of performance he wanted was possible. The basic description of the CL-400-15JP (a crew of two in a vehicle with a gross weight of 150,000lb with two J58 turbojets) turned out to be incredibly prescient, though the actual configuration bore little relation to what would come later.

Beginning in April 1958, Johnson started looking seriously at a Mach 3 hydrocarbon-fuelled recon platform. While inspired by the potential of the CL-400-15JP, his new concepts bore little similarity to that design. Study of Soviet radar systems capability had indicated what would be required of an aircraft to avoid detection: flight of Mach 3 or more above 90,000ft, coupled with a radar cross section of ten square metres or less. Such an aircraft would spend too little time in radar beams, and return too little of a signal, to be reliably detected. Johnson penciled a vague back-of-the-envelope notion dubbed 'U-3' in April 1958. Rough notional concepts evolved into a more detailed design in June of 1958, a design nicknamed 'Archangel'.

The Archangel and its descendants led directly to the CIA's A-12 and the SR-71. The path from beginning to end is incompletely known, with some of it classified, some likely lost, and a fair amount simply squirreled away and as yet unrevealed. But there is enough available to work out a good summary of how the Lockheed Blackbirds evolved. Rather than being derived from space alien technology, or springing fully formed out of the mind of a design genius, the Blackbirds evolved through a series of designs, much like any other aircraft throughout history. The end result was amazing for its – or indeed our – time, but its development was not miraculous, simply the result of hard work by many smart people over a period of years.

Archangel was still a very preliminary design, but it was a start. The math showed that the concept was feasible and could reach CIA performance goals. It was a fairly conventional 1950s supersonic aircraft configuration... swept shoulder wings, pointed fuselage, underslung engines tucked in the corner between wings and fuselage, and a highly swept tail unit that appeared to be just stuck on. The tail was the most unconventional feature (it looks almost like a simplistic afterthought), and was the design element that would change the fastest as the Archangel design became the Archangel series. Johnson believed that the design could be delivered 18 to 24 months from go-ahead, a view that today would brand a designer a fevered madman... but for Kelly Johnson in the late 1950s, it was entirely practical.

Archangel I

The Archangel was a very preliminary design featuring a vertical tail that was simply unworkable. So the design was literally taken back to the drawing board, and a week later a more refined design emerged. This, the Archangel 1 (or A-1), was the first true serious design in the long series that led to the SR-71.

The A-1 was an attractive single-seat titanium (B-120VCA, an alloy with 13% vanadium 11% chromium and 4% aluminium) aircraft equipped with high shoulder-mounted wings and a J58 turbojet tucked beneath each of them, next to the fuselage. The configuration was much like that of the original Archangel, but the tacked-on tail of the original was now a more realistic appendage attached to a slightly upswept aft fuselage. The design work included a detailed inboard profile. To improve altitude performance, large ramjets could be added to the wingtips... but the added weight increased the wing loading and reduced range.

Kites

A diversion occurred in August of 1958. The Navy was also interested in high-speed, high-altitude reconnaissance aircraft, toying with a Goodyear concept of a balloon-launched inflatable supersonic aircraft. This was doubtless related to the Convair semi-inflatable MC-10 aircraft concept, resembling it in many respects. Lockheed was asked to weigh in on the idea and Johnson took a dim view of it. In order to lift the heavy aircraft high enough in the sky, he estimated the balloon would end up being a mile in diameter. This would be extremely impractical.

Nevertheless, from August to September 1958 the Skunk Works considered a few aircraft concepts that

Lockheed 'Archangel'

SCALE 1/200

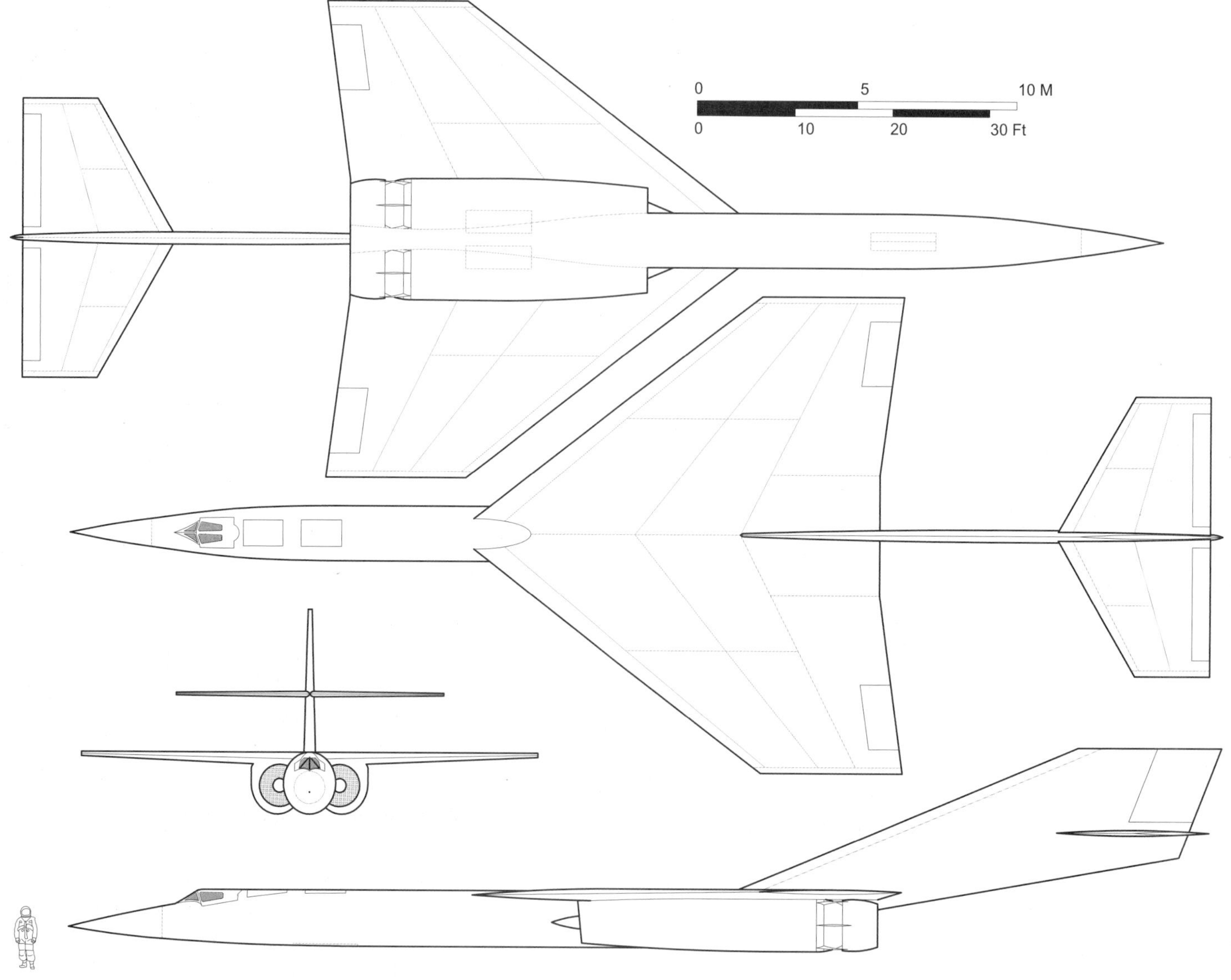

were similar to the MC-10 in configuration, if not in materials and launch method. Lockheed looked at inflatable concepts as well as balloon and rocket launched designs, but the only ones that have so far come to light are two metal-structure designs with large ramjets. Dubbed 'Ram Jet Kites', these designs shared a large 15ft-diameter ramjet that served as the main fuselage, the cockpit located in the inlet centrebody cone. One design suspended the ramjet below a large swept wing with vertical tip fins, while the other mounted the wings at the midpoint of the ramjet nacelle. As with the MC-10, the Kites were to use borane-based fuels.

Available data on the 'Kites' is lean. However, included with the 'Kite' concept was the need for a tow aircraft. Instead of a vast balloon, or carrying the aircraft on the back of a B-52, the 'Kites' were to be towed to altitude by a specially designed aircraft. Known as the 'Peterbilt' after the cargo-hauling truck brand, this was a large four-turbojet subsonic aircraft meant to do only one thing: pull a Kite to 70,000ft and Mach 0.8.

The design was straightforward, with no military systems or concessions towards reduced radar cross section. It was merely a tow truck of the skies, looking vaguely like a mutant twin-boom U-2. After separation from the Peterbilt, the Kite would use a rocket system to accelerate to ramjet speed and climb to a higher altitude.

Unfortunately for the concept, while Johnson's team decided that the towed metal-structure aircraft would be perhaps 80% the weight of the competing inflatable designs, they still could not quite meet the intended performance. The Peterbilt and the Kites seem to have been a flash in the pan, soon dropped and forgotten.

Archangel II

The Archangel II design was intended to improve on the Archangel I's altitude performance. It had a similar

Lockheed A-1

SCALE 1/185

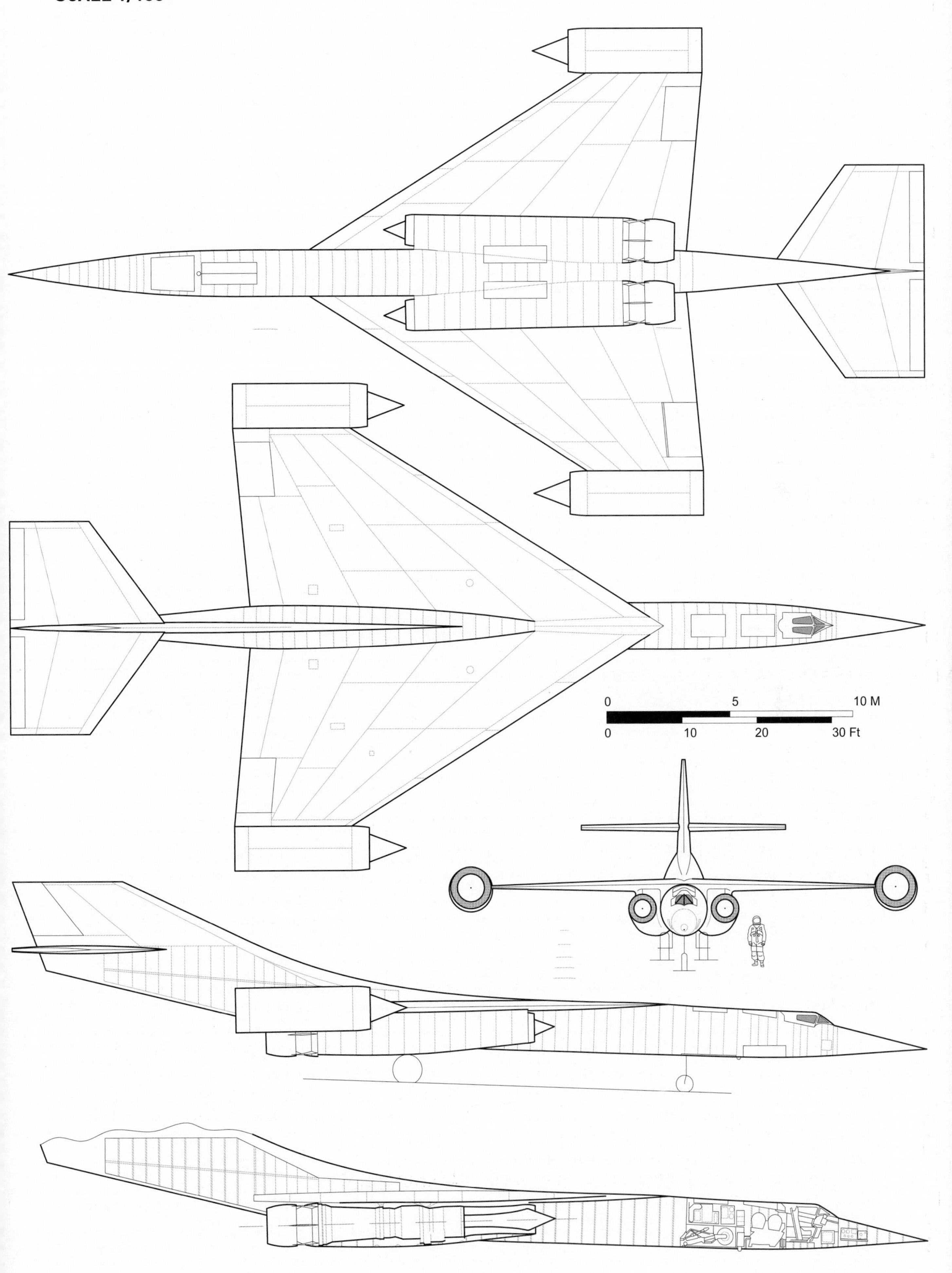

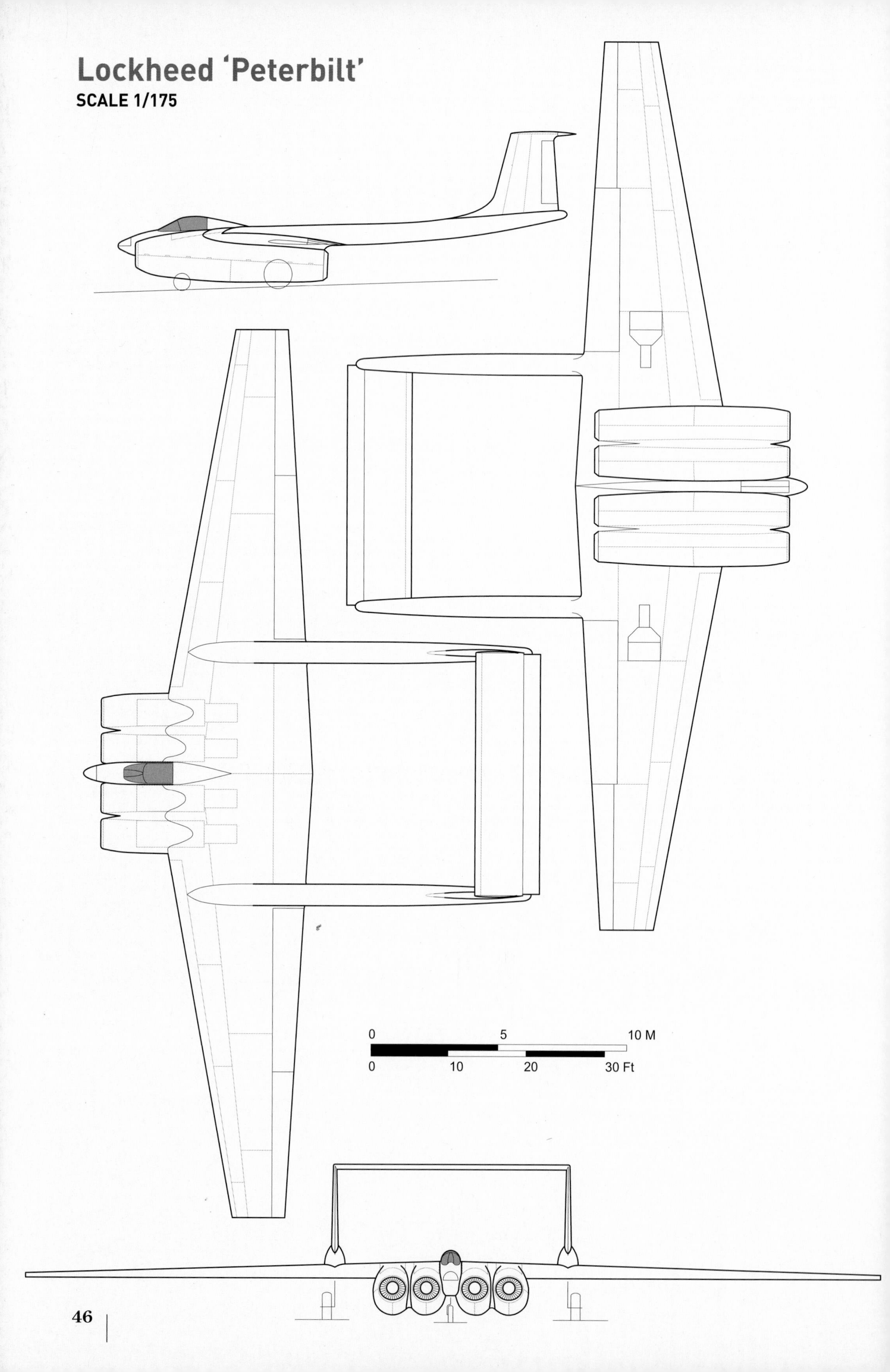
Lockheed 'Peterbilt'
SCALE 1/175
0
5
10 M
0
10
20
30 Ft

Lockheed 'Kite' High Wing

Panel Lines Speculative

SCALE 1/400

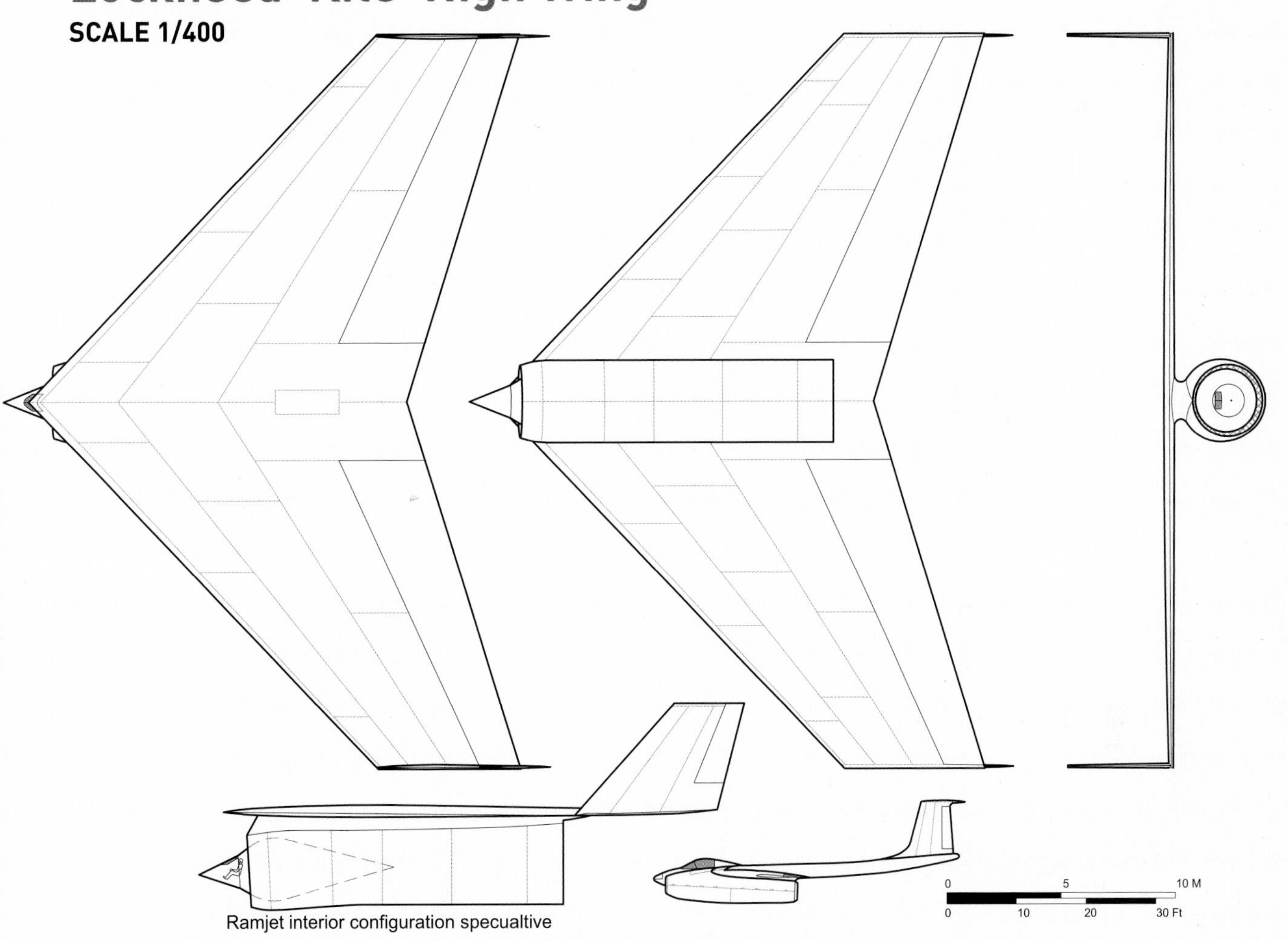

Ramjet interior configuration specualtive

Lockheed 'Kite' Mid Wing

SCALE 1/400

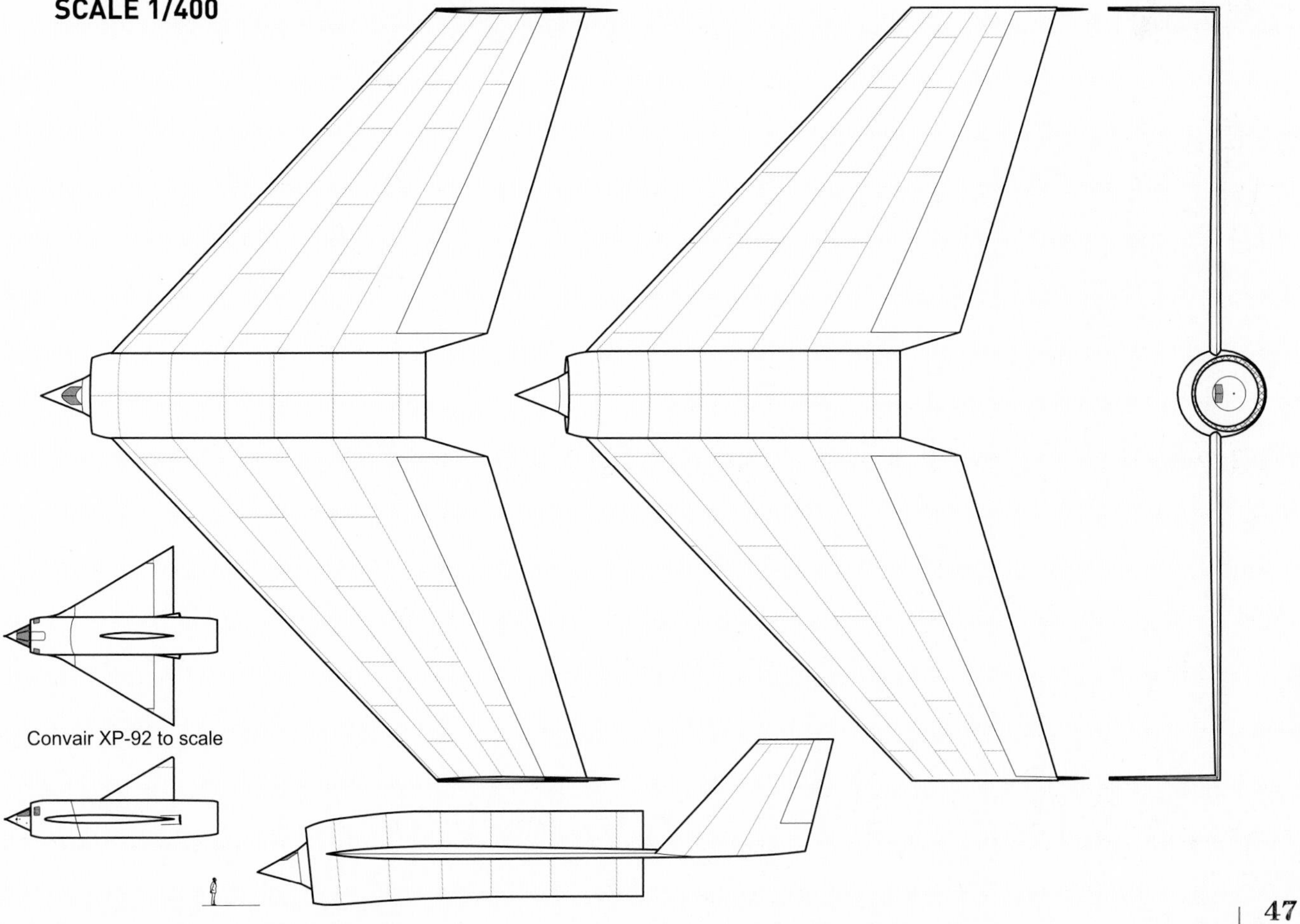

Convair XP-92 to scale

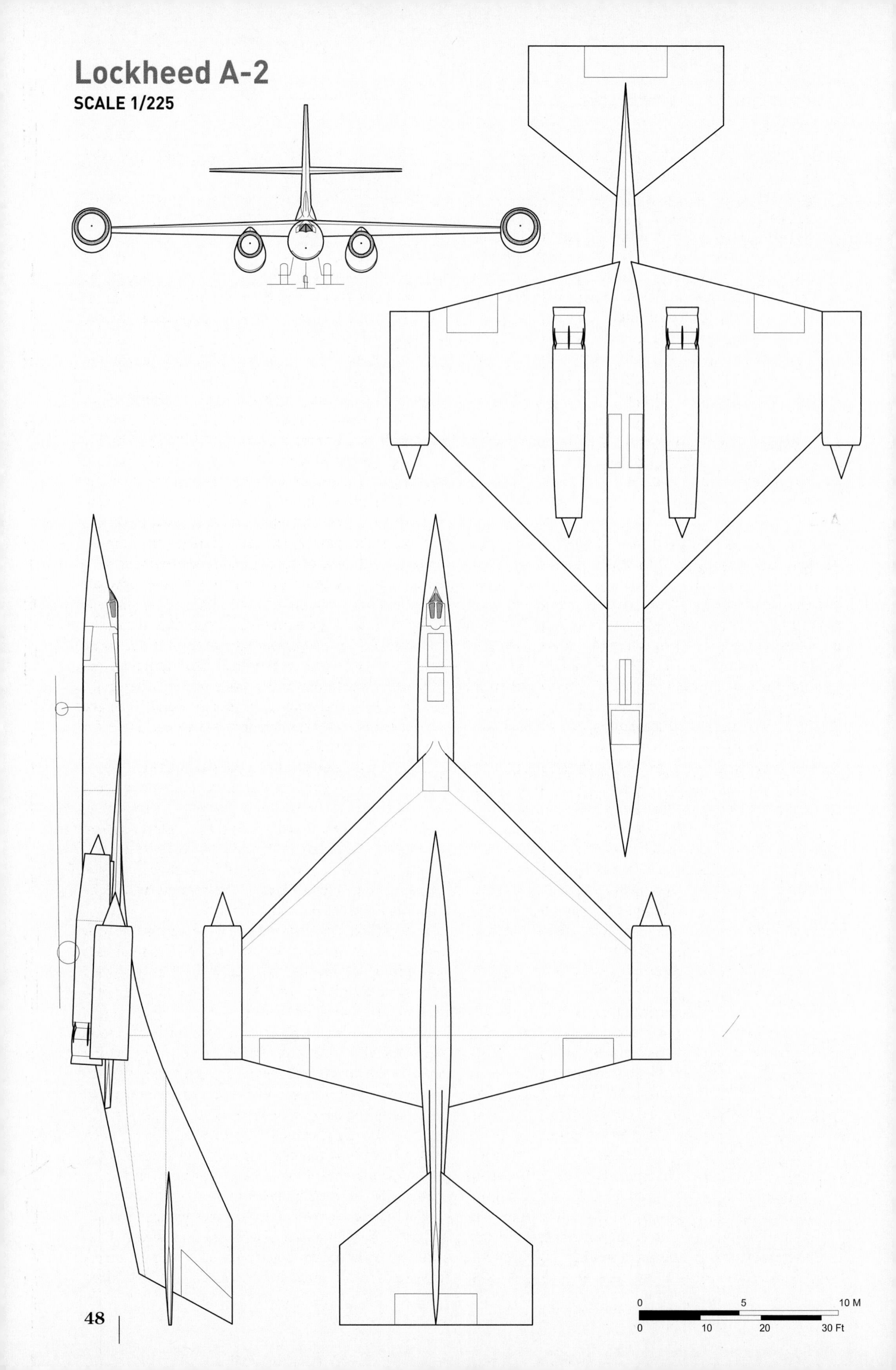

Lockheed A-2

SCALE 1/225

configuration, with a similar gross weight, but the areas of the wings and tail were substantially increased and the 75in wingtip ramjets were permanent. It would also be a dual-fuel vehicle, with the ramjets burning pentaborane and the J58s burning JP-150 (a kerosene fuel derived from JP-4). The J58s were moved outboard, away from the fuselage, solving a problem of turbulent airflow into the Archangel I's tucked-in inlets. Moving the turbojets outboard also reduced the moment arm on the wings, lowering wing weight.

The problems with pentaborane were not the downfall of the Archangel II design. Instead, the need was for a stealthy aircraft nearly invisible to radar, and no concessions to reduced radar cross section had been made. The Archangel II was a large aircraft with many sharp corners – perfect for reflecting radar signals right back to the sender.

A-3

The next step that the Skunk Works took was to design a much smaller vehicle with some effort made at radar cross section reduction… much of it due to simply being a smaller target. The 'Archangel' designation seems to start getting truncated at this point, resulting in this design series being dubbed the A-3. Instead of the J58, Johnson looked at the Pratt & Whitney JT12, a much smaller engine that found use on such subsonic aircraft as the Lockheed Jetstar and XV-4 Hummingbird and the North American T28 Buckeye.

As a relatively low pressure engine, he thought that it might work well for high Mach applications. So by assuming a JT12A engine with afterburners coupled with separate ramjet engines, he penciled out a few very preliminary concepts including tailless and semi-flying wing concepts. In October of 1958 detailed design work for the new 'Archangel 3' was turned over to other designers at the Skunk Works who created at least five concepts for similar small designs. As these 'Archangel' concepts were much smaller than what had come before, they were initially dubbed 'Cherubs'.

Some basic information is available on several 'Cherub' concepts, though little more than what's on the original general arrangement diagrams. The goals for the Cherub included the use of two JT12A turbojets with afterburners; two ramjets (potentially borane-burning); initial cruise altitude of 90,000ft with altitude over the target at 95,000ft; radius at Mach 3 to 3.2 of 1500 nautical miles with jet fuel and 2000 nautical miles with borane; and an empty weight of around 9000lb.

Cherub I, designed by Ed Baldwin, was a stubby-looking tailless clipped delta concept with turbojets in the fuselage fed by 2D ramp intakes. The ramjets were in flattened wingtip pods and added fuel was stored in underwing drop tanks. The elliptical planform of the ramjet intakes was an attempt at reducing radar return though the large slab-sided vertical stabilizer and rectangular inlets for the turbojets would have been bright radar reflectors.

Cherub 2, designed by Dan Zuck, was a sleeker-looking beast than the Cherub 1, with smaller, more sharply-swept wings and turbojets fed by low-mounted circular inlets. Otherwise the configuration was much like the Cherub 1. The 'Cherub Variant' by Henry Combs was basically the Cherub 1, but with the turbojets moved from the fuselage to nacelles located over the wings at about one-third span. This would have improved airflow for the inlets and potentially reduced the weight of the wing structure.

The 'A-3 Variant', also designed by Zuck, looks like it would not have been out of place in a Star Wars movie. It was almost a flying wing, with a highly swept wing capped with a long pointed nose, a short tail and a surprisingly conventional cockpit canopy rising from above the thickened wing centre section. The turbojets were embedded within the wing section, flanking what passed for the fuselage; the ramjets were just outboard of the turbojets, also embedded in the wings. The exhaust nozzles for both the turbojets and ramjets were to be made of 'plastic', in this case doubtless meaning a high temperature fibreglass.

The final A-3 design was derived from the 'Cherub Variant' and had the same overall configuration, but with wings much more like those of the 'Cherub 2'. The final A-3 was studied in greater detail than the other designs and was put forward to the DoD for consideration in a November 1958 proposal.

The wingtip ramjets were not purely dead weight during the early climb phase. They were configured as jet fuel tanks, able to hold 1100lb of jet fuel each. It was thought that to reduce drag during this stage of the flight the aft ends of the ramjets could be capped off with conical Mylar 'sacks' held to shape with air pressure. The inlet spikes of the ramjets would be movable to obtain optimum thrust.

The cockpit was reduced in size compared to that of the U-2. It would doubtless be claustrophobic and uncomfortable, but the mission time was expected to only be a fifth that of a U-2 mission. The need to save weight was taken to an extreme in the cockpit… flight instruments (such as rate of climb indicators) were deleted where possible; the controls stick and pedals were to be redesigned for minimum weight. The ejector seat would only be used on training and ferrying flights; on operational missions the pilot would have a lightweight non-ejecting seat. The pilot would wear a Navy-type full pressure suit, important because at altitude the cockpit would be pressurized with pure nitrogen and air conditioned via boiling off water.

The landing gear was of tricycle arrangement, but was so narrow track as to seem like bicycle. For takeoff, U-2-style 'pogo' stabilizers would keep the wingtip ramjets from striking the ground; after landing, it was hoped

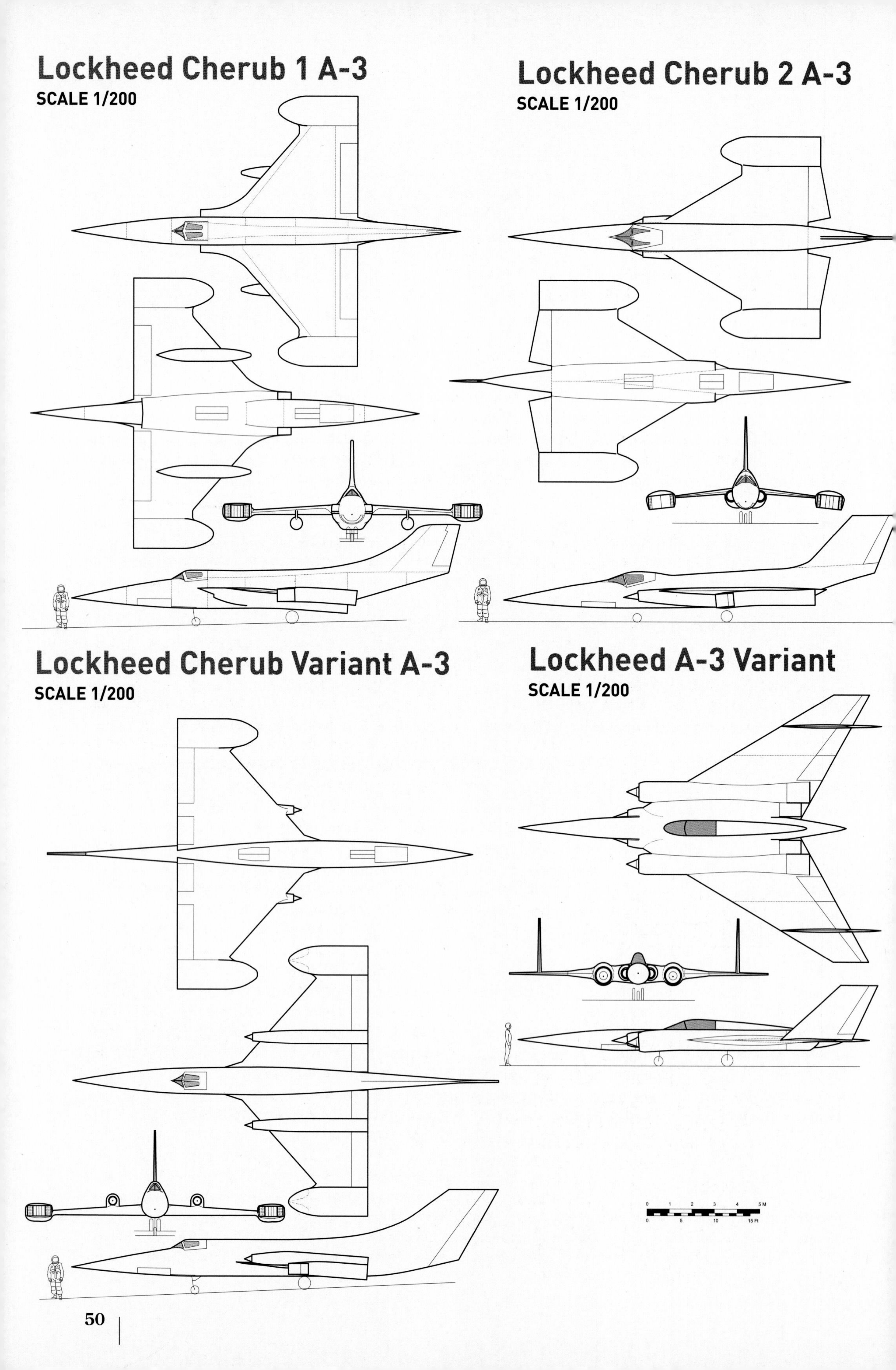
Lockheed Cherub 1 A-3
SCALE 1/200
Lockheed Cherub 2 A-3
SCALE 1/200
Lockheed Cherub Variant A-3
SCALE 1/200
Lockheed A-3 Variant
SCALE 1/200
0 1 2 3 4 5 M
0 5 10 15 Ft

that the plane would retain stability long enough so that when it did tip over it would not damage the ramjets.

The structure would be largely of titanium, fabricated from sheets of .016 gauge, down to 0.010 gauge where possible. Use of plastic (i.e. fibreglass) was not expected, though that would be revised if RCS testing showed it necessary for the inlets or the tail. Testing on 1/40 and 1/20 scale models had shown that the average radar return of the A-3 was about that of the U-2, but with higher spikes driven by the ramjets and the fuselage. The turbojet nacelles were merged into the upper surface of the wing leaving an unbroken lower surface for reduced radar reflectivity.

It was planned to use petroleum-based JP-150 jet fuel. If high energy borane fuels became available for the ramjets at a later time, the large fuel tanks designed for the jet fuel would provide more than enough volume for much improved range performance... going from a mission radius of 1500 nautical miles to 2020 nautical miles. If the A-3 was refuelled at altitude and speed from another A-3, the mission radius would be 2050 nautical miles using JP-150 only. Turnaround time was to be three to four hours, with two sorties per day.

In November of 1958, the Air Force compared the A-3 with the Convair FISH and concluded that FISH was the superior option. Lockheed's design came out ahead in operational and logistical metrics, but the Convair design simply performed better. It was time to go back to the drawing board.

A-4

Johnson started from scratch again in late 1958, this time with an eye towards greatly improved radar cross section. The A-4 started a series of designs with some common features, most notably radar-defeating chines and blending. The A-4 of December, 1958, had a single afterburning J58 turbojet, fed by an underslung semi-circular inlet. Its most prominent feature was an oversized vertical stabilizer that reached forward almost all the way to the cockpit; it was heavily blended into the fuselage, giving little flat area for a large radar reflection spike.

The A-4-1 and A-4-2 (drafted by Ed Baldwin) differed in that the -2 configuration had 34in ramjets added to the wingtips to aid in increasing altitude. But as had been found with Archangel I, the addition of ramjets harmed range. This ended the brief existence of the A-4, since the range of the A-4-1 was already 240 miles short of the goal.

A-5

The A-5 of December 1958 (also drafted by Ed Baldwin) was one of the smallest of the Archangel designs. Designed at the same time as the A-4, it used two of the smaller JT-12A engines in conjunction with a single large (82in exit diameter) ramjet with a configuration roughly like that of the A-4, including the large, blended vertical stabilizer. But very much unlike the A-4 – and all other Archangel designs – the A-5 was also equipped with an Aerojet hydrogen peroxide rocket engine (available documentation is unclear whether the engine was a monopropellant rocket engine, or a bipropellant engine burning hydrogen peroxide with jet fuel).

The rocket was to be used to boost the plane from the highest altitude the small JT-12As could give to the speed and altitude needed for the ramjet to operate. Given the inadequate range performance coupled with the complex composite propulsion system, it's unsurprising that the original diagram for the vehicle was incomplete.

A-6

The next in the series, the A-6, received a bit more study. At least nine configurations were created from December 1958 into January 1959; diagrams and information on three of them (all drafted by Ed Zuck) have come to light. The three are similarly configured, with a cranked delta configuration, a single Pratt & Whitney J58 and two ramjets. In all three cases the J58 was in the centre-rear of the fuselage fed by a semi-circular intake under the nose with a long duct; the ramjets flanked the turbojet and were fed from separate semi-circular intakes under the middle to rear of the fuselage. All three also feature extensive blending and wide chines that lead from the wing leading edges to the nose.

The A-6-5 had two vertical stabilizers at about mid span, canted inboard to reduce radar reflection. The A-6-6 has similar stabilizers, wider chines and an added vertical stabilizer. This latter feature was somewhat like that of the A-4 and A-5, with a wide base but without the extensive blending. The A-6-9 did not have the central stabilizer, but did tip up the outer wing panels outboard of the vertical stabilizers. The vertical stabilizers were located somewhat further inboard than on the earlier designs, and had forward-swept trailing edges. The J58 inlet was moved aft, almost to mid-fuselage. This necessitated a distinctly serpentine inlet duct... unfortunate from an aerodynamic standpoint, but a bonus from the radar cross section point of view since the face of the turbine could no longer be directly seen through the intake.

Arrow and B-58 Launched Vehicle

Also in December came a minor diversion in effort. As a check on whether the two-stage Convair FISH concept actually made sense, Lockheed was asked to design similar aircraft. Little is available on the

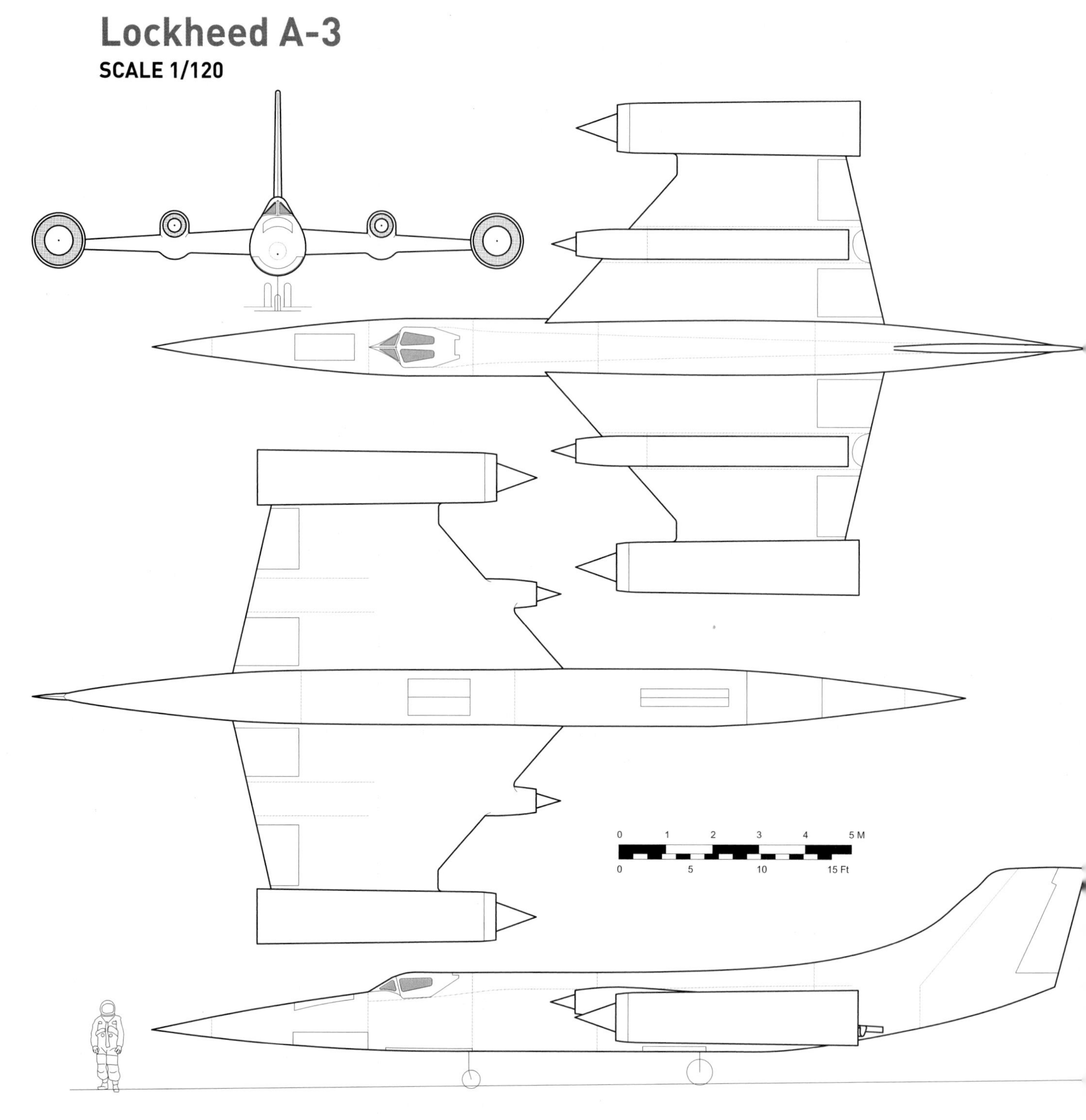

two designs that resulted apart from a few diagrams and photos of a model. One design, dubbed Arrow I, was drafted by Dan Zuck in early December. Appropriately, it looked like a sharply swept arrowhead with an underslung inlet and two mid-span vertical stabilizers, tipped very slightly inboard.

The other design, the mid-December 'B-58 Launched Vehicle' by Henry Combs, was somewhat similar but less highly swept and slightly more rotund. Both designs featured two 40in ramjets and a single P&W JT12 turbojets for landing. Both would be carried to altitude and speed by a B-58. In order to fit the bomber's landing gear, the 'B-58 Launched Vehicle' had 'doors' in its wings that the main gear could pass through. One had a range of 1736 nautical miles, the other 2208… but it's not clear which one was which. The two designs validated the possible performance of the Convair FISH but also showed the difficulties with the design, such as the inability of the pilot to eject while carried by the B-58.

A-7

While the A-3 through A-6 had produced small designs with increasingly successful RCS reduction,

Lockheed A-4-2

SCALE 1/144

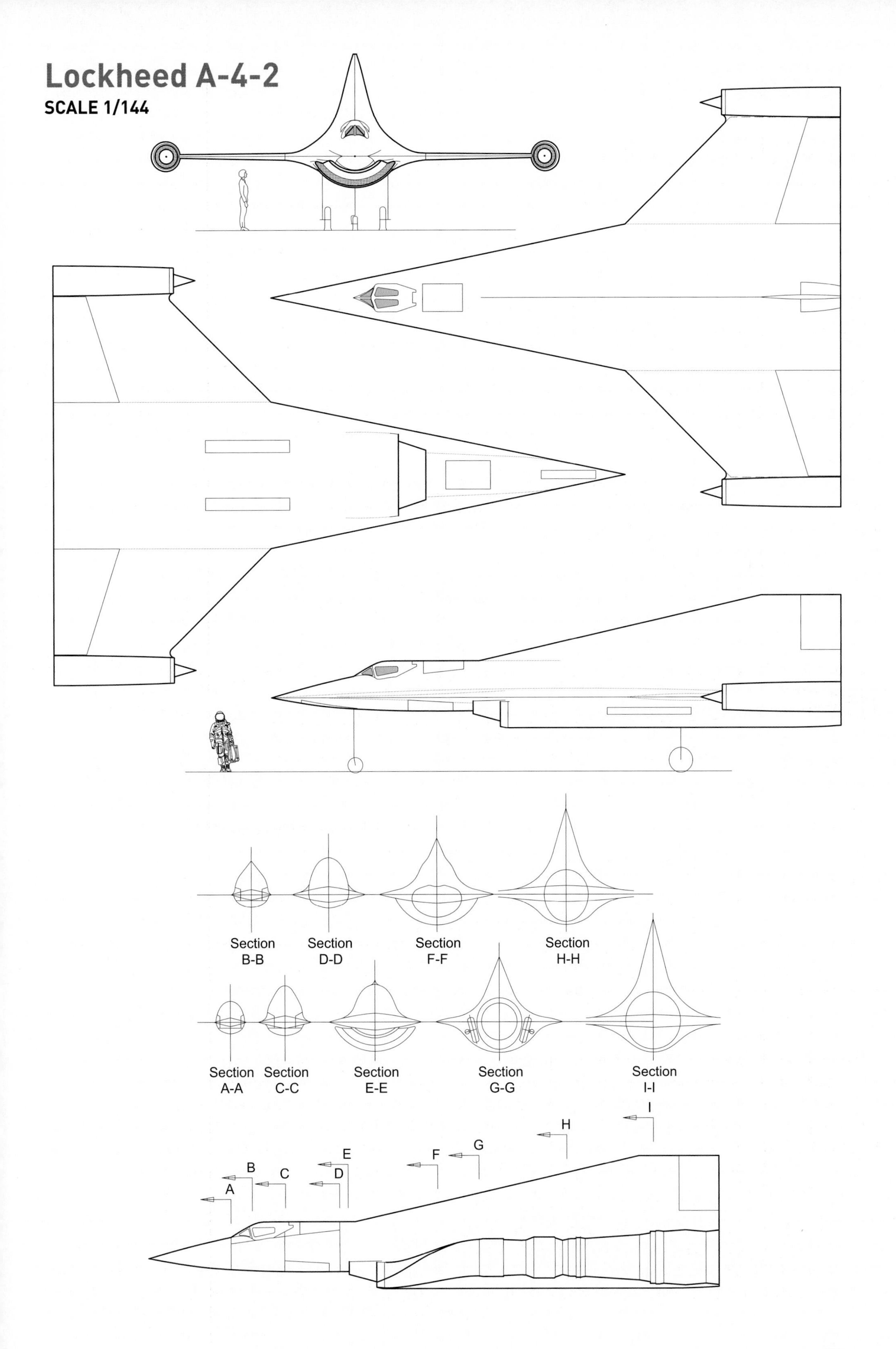

Lockheed A-5

SCALE 1/125

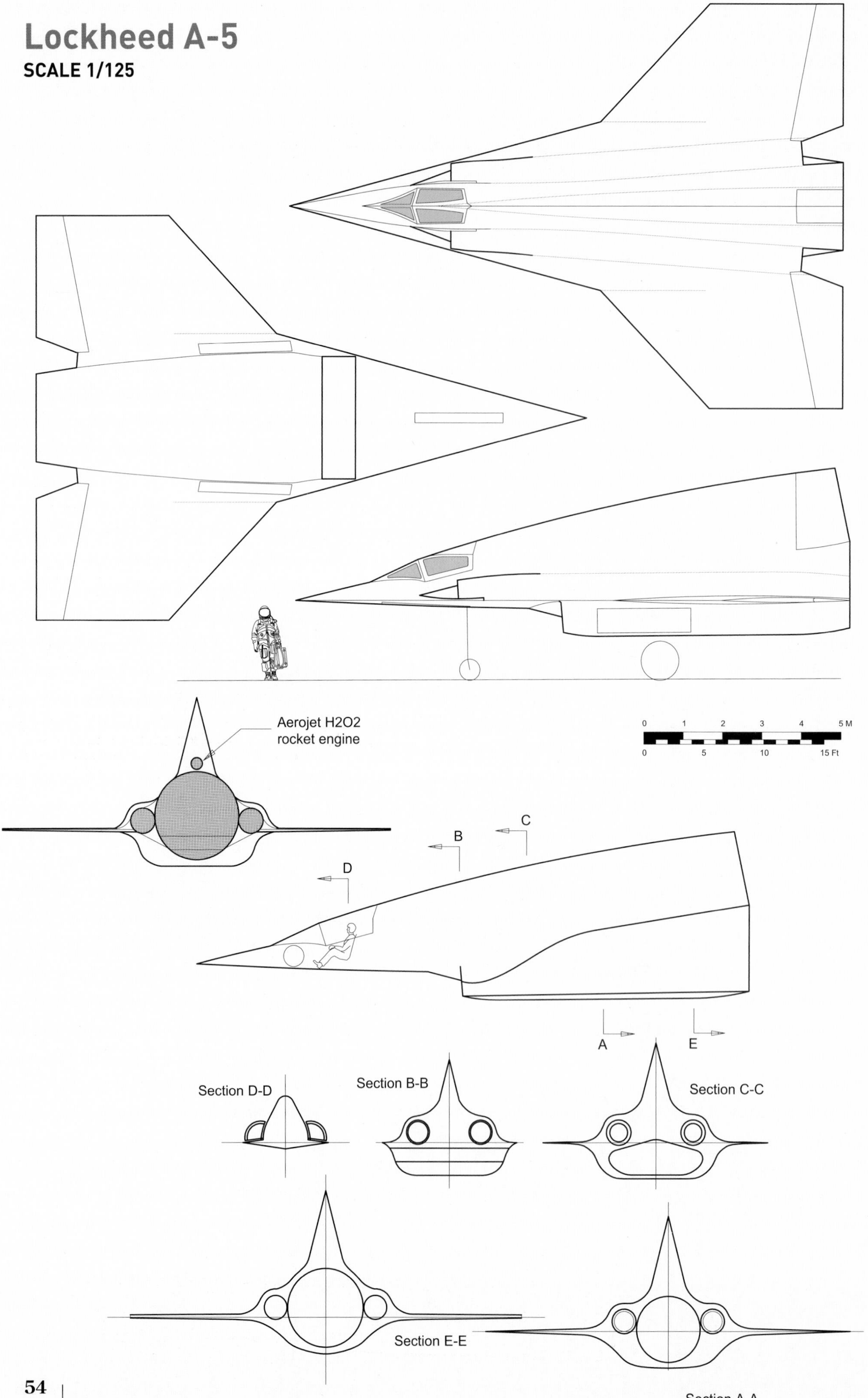

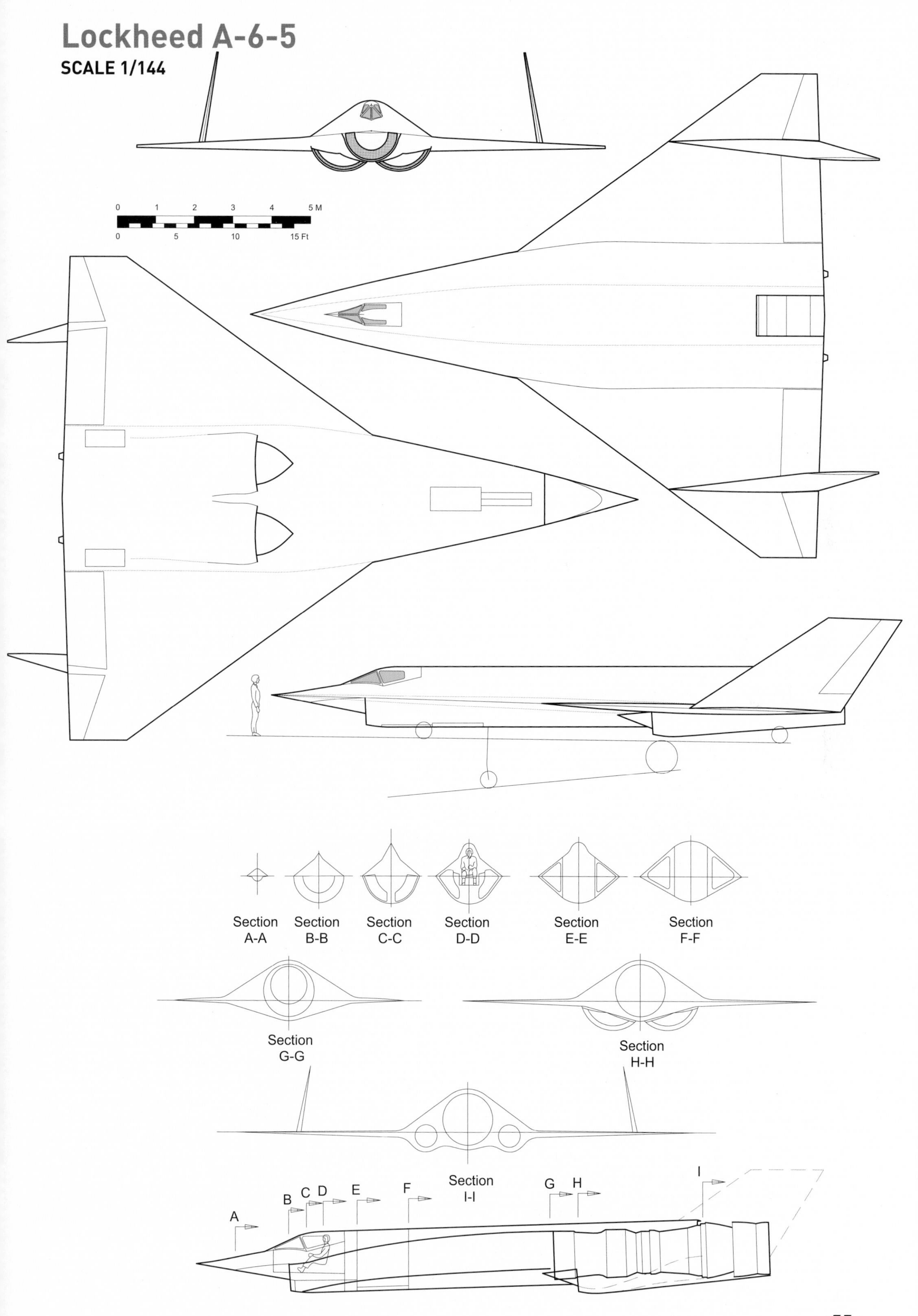

Lockheed A-6-5
SCALE 1/144
0 1 2 3 4 5 M
0 5 10 15 Ft
Section A-A
Section B-B
Section C-C
Section D-D
Section E-E
Section F-F
Section G-G
Section H-H
Section I-I
A
B
C
D
E
F
G
H
I

Lockheed A-6-6

SCALE 1/130

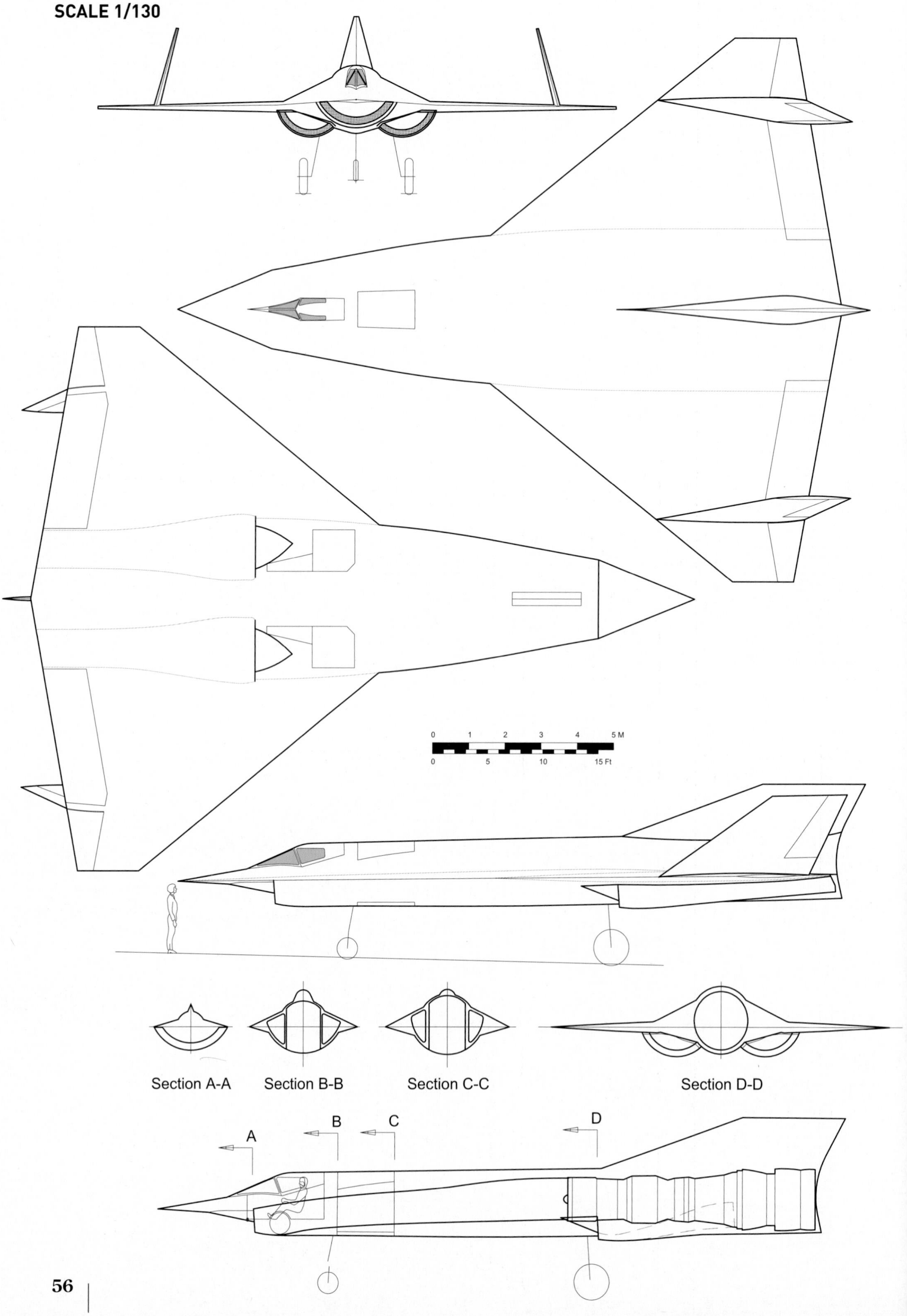

0 1 2 3 4 5 M

0 5 10 15 Ft

Section A-A

Section B-B

Section C-C

Section D-D

A

B

C

D

Lockheed A-6-9

SCALE 1/170

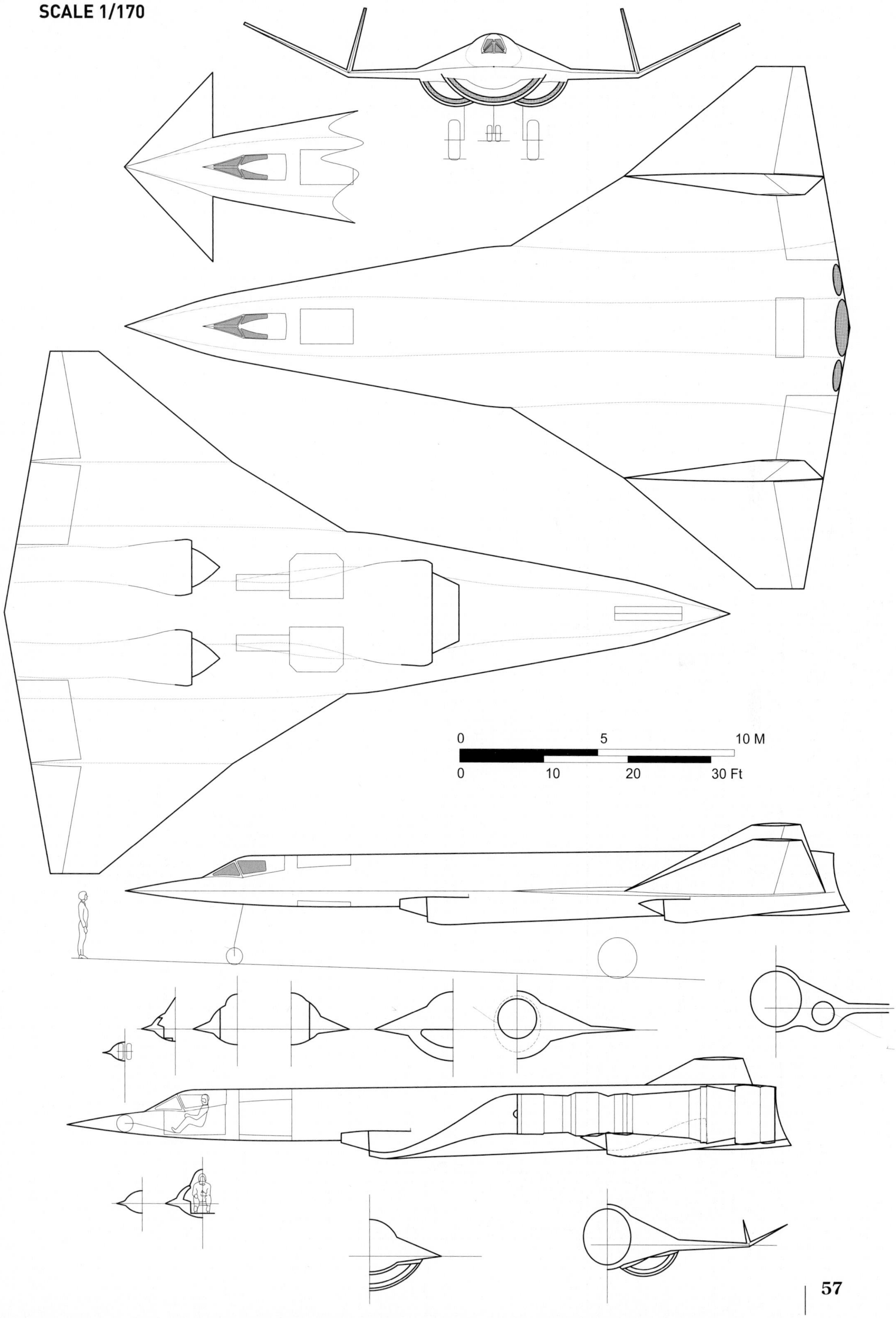

0 5 10 M

0 10 20 30 Ft

Lockheed Arrow

SCALE 1/100

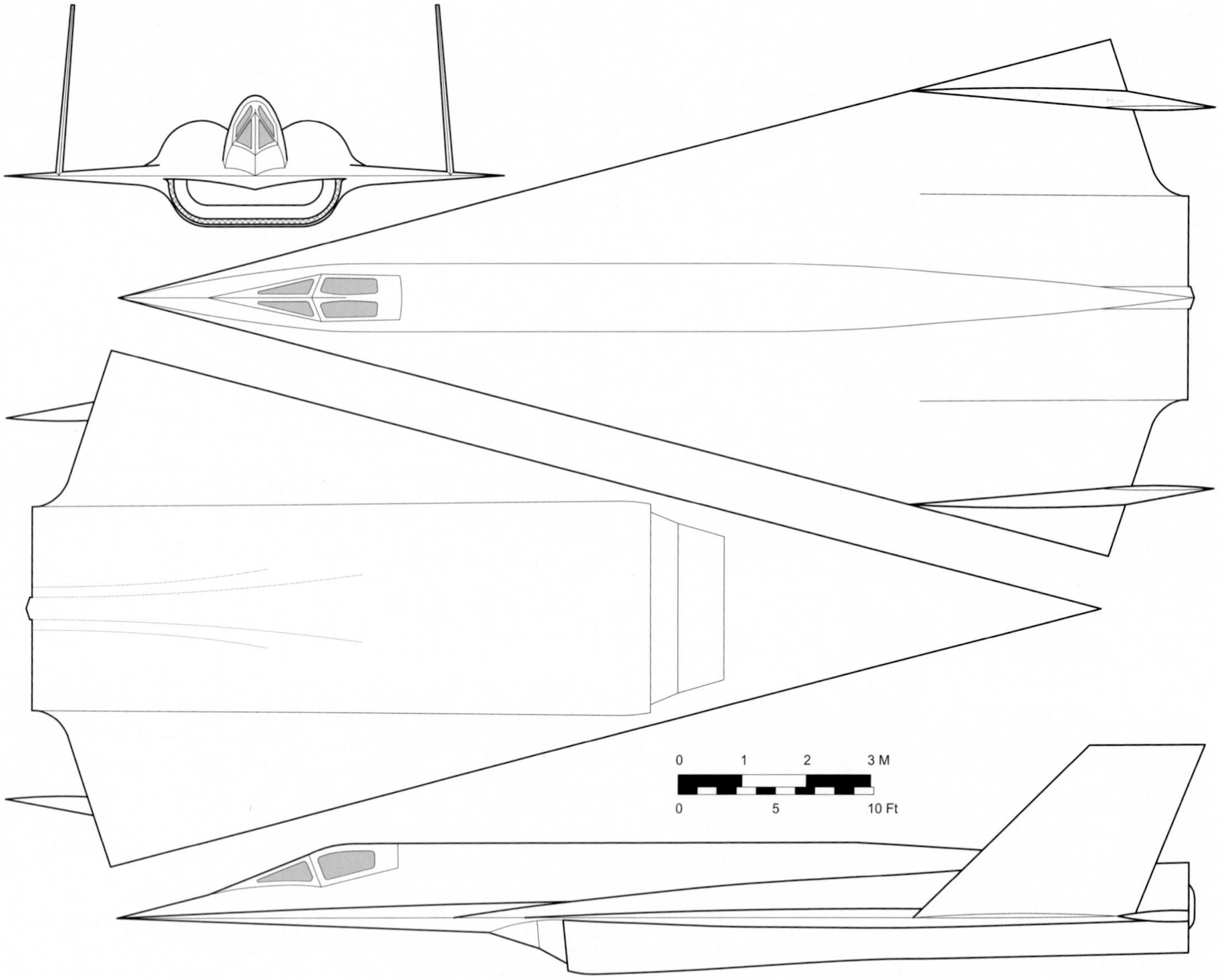

0 1 2 3 M

0 5 10 Ft

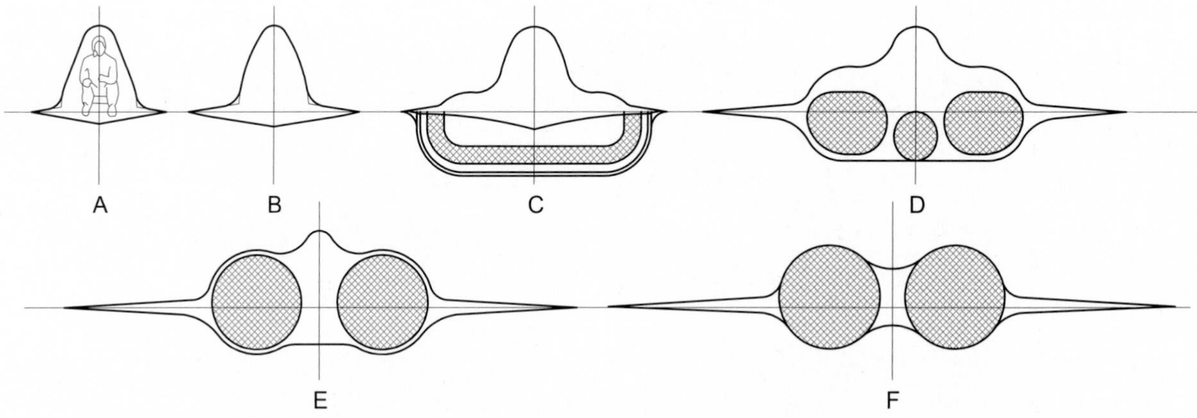

A B C D

E F

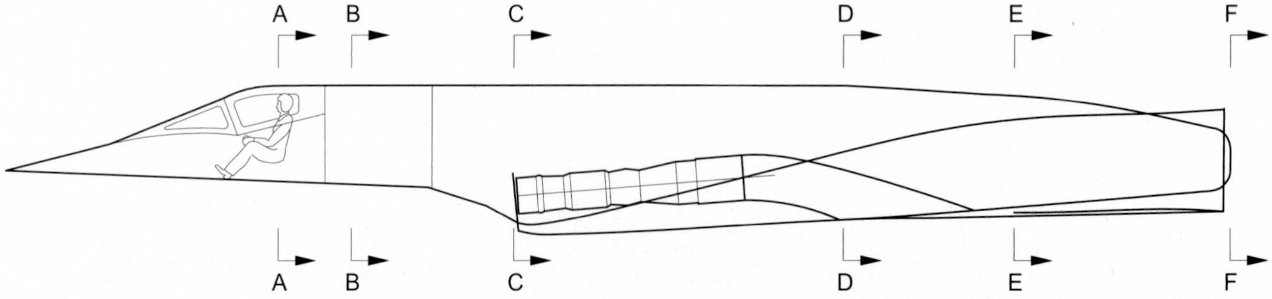

A B C D E F

A B C D E F

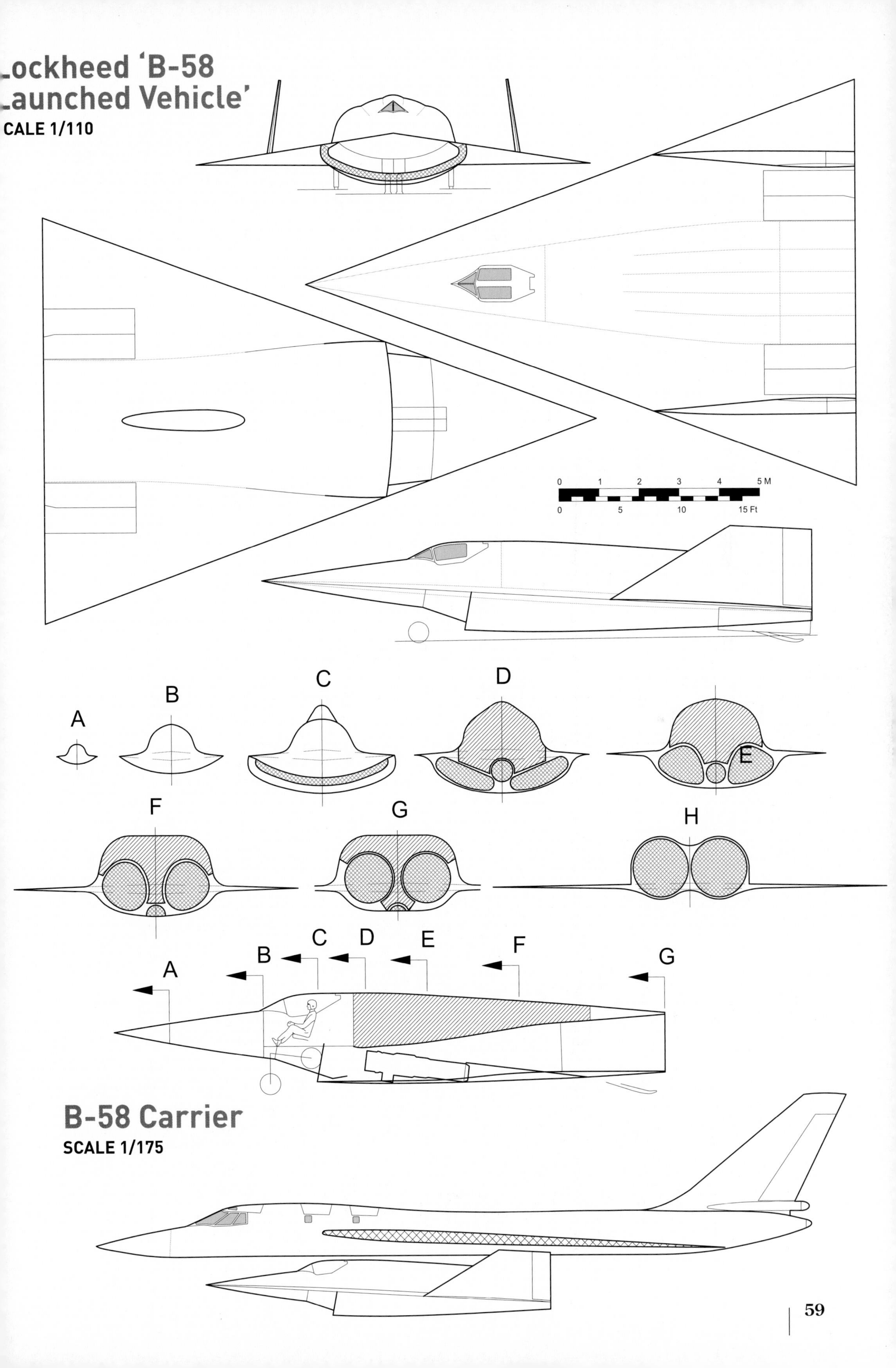
Lockheed 'B-58
Launched Vehicle'
SCALE 1/110
0 1 2 3 4 5 M
0 5 10 15 Ft
A
B
C
D
E
F
G
H
B-58 Carrier
SCALE 1/175

the actual performance of the aircraft in terms of range and altitude never quite measured up. The combination of ramjets and turbojets was turning out to not be the panacea that had been hoped, especially when compounded with the aerodynamic concessions required for low radar cross section. So in January 1959, Kelly Johnson again changed course, this time examining designs with no RCS concessions. The A-7 series, for which two diagrams by Ed Baldwin are available, returned to the original Archangel I concept of fast uncompromised aircraft.

The A-7-1 and A-7-2 are apparently much the same design except that the A-7-2 had ramjets attached to the wingtips. To some degree they were conventional aircraft laid out much like Archangel I, but with two major differences. The first was that the A-7 had a single turbojet rather than the Archangel I's two... and its wings were set so low that they were actually below the fuselage. The leading apex of the diamond planform wings touched the underside of the fuselage, but due to 4° of incidence, the blunt trailing edge of the wing was well below the aft fuselage. This provided a space for the exhaust nozzle of the J58 turbojet, while also providing a fairly smooth and unbroken lower surface. The end result is an unusual and uncomfortable-looking aircraft that the designer himself found unlovely.

The A-7-3 was somewhat scaled up, but more importantly moved the wing to the top of the fuselage while keeping 4° of incidence. This resulted in an overall more conventional looking aircraft even though the individual components were much the same.

Little is available on the performance of the A-7 designs apart from a typical mission radius of 1637 nautical miles. In February of 1959 some effort was made to study the use of inflight refuelling to extend the range of the A-7... supersonic refuelling using a modified A-7 as a tanker. Unsurprisingly this proved to be impractical.

A-10

Unfortunately, little is known of the A-8 and A-9 apart from the fact that they, like the A-7, used a single J58 and two ramjets and were thus relatively small aircraft. Repeated attempts to design a small aircraft failed to produce a design that could attain the desired range. So once again Johnson changed directions. A large aircraft with multiple turbojets and no attempt at RCS reduction became the A-10.

This aircraft was geometrically fairly simple – a cylindrical fuselage with slim nose and tail, a single large vertical stabilizer, shoulder-mounted diamond wings and turbojets in individual underslung nacelles. With the A-10, the turbojets used were the J93s designed for the North American B-70. At this time the J93 was thought to be ahead of the J58 in development, a belief that turned out not to be correct.

The A-10 was designed to go high, fast and far; it was not designed to do it stealthily. Nevertheless, some effort was made to study a reduced RCS version that used additional flat panels added to a scale model. This had the effect of adding chines to the fuselage and nose, and somewhat blending the cylindrical engine nacelles into the wing. The results are not known but doubtless this modification would have substantially increased drag and weight.

A-11

The A-10 came close. Close enough, in fact, that Johnson felt fairly confident that the next design, the A-11, would be the final configuration. Consequently the A-11 of March 1959 was designed to a fairly high level of detail.

It was quite similar to the A-10 in configuration, with no major changes. The only noticeable alteration was to un-clip the wingtips. The A-11 was to be built largely of B-120VCA titanium. The outer skin of the wing was fabricated from 0.020in B-120VCA titanium cold roll sheet, spot welded to an inner skin made from 0.025in B-120VCA. The outer skin was a smooth surface while the inner skin was corrugated. As the skin heated during high-speed flight, the outer surface would buckle and warp, but it would be held in check by the inner skin. Those inner corrugations would force any buckling in the outer skin to form in the streamwise direction, thus not fundamentally altering the aerodynamics of the wing.

The inner structure of the wing was formed by a titanium box from 15% to 80% of the wing chord, running to just outboard of the nacelles. The outer wing panels were designed to be removable to facilitate maintenance, replacement and transportation. The wing box was fabricated from ribs every 40in and beam every 16in. While fairly thin, the wings nevertheless held a substantial fuel volume due to having a blunt trailing edge, an unusual design choice.

The fuselage was of semi-monocoque construction of titanium skins (0.016in thick at the nose, increasing to 0.040in in the centre section), rings and four longerons. It was cylindrical except for long tapering nose and tail cones. The nose did not taper smoothly though, having a stepped appearance in planform.

Behind the cockpit was the sizable military equipment bay. It was 72in long and accessed through top and bottom doors 58in long. The 4ft-long bottom doors would be used for equipment loading and unloading of the 500lb payload.

The A-11 had, compared to many prior designs, a much simplified propulsion system with only two

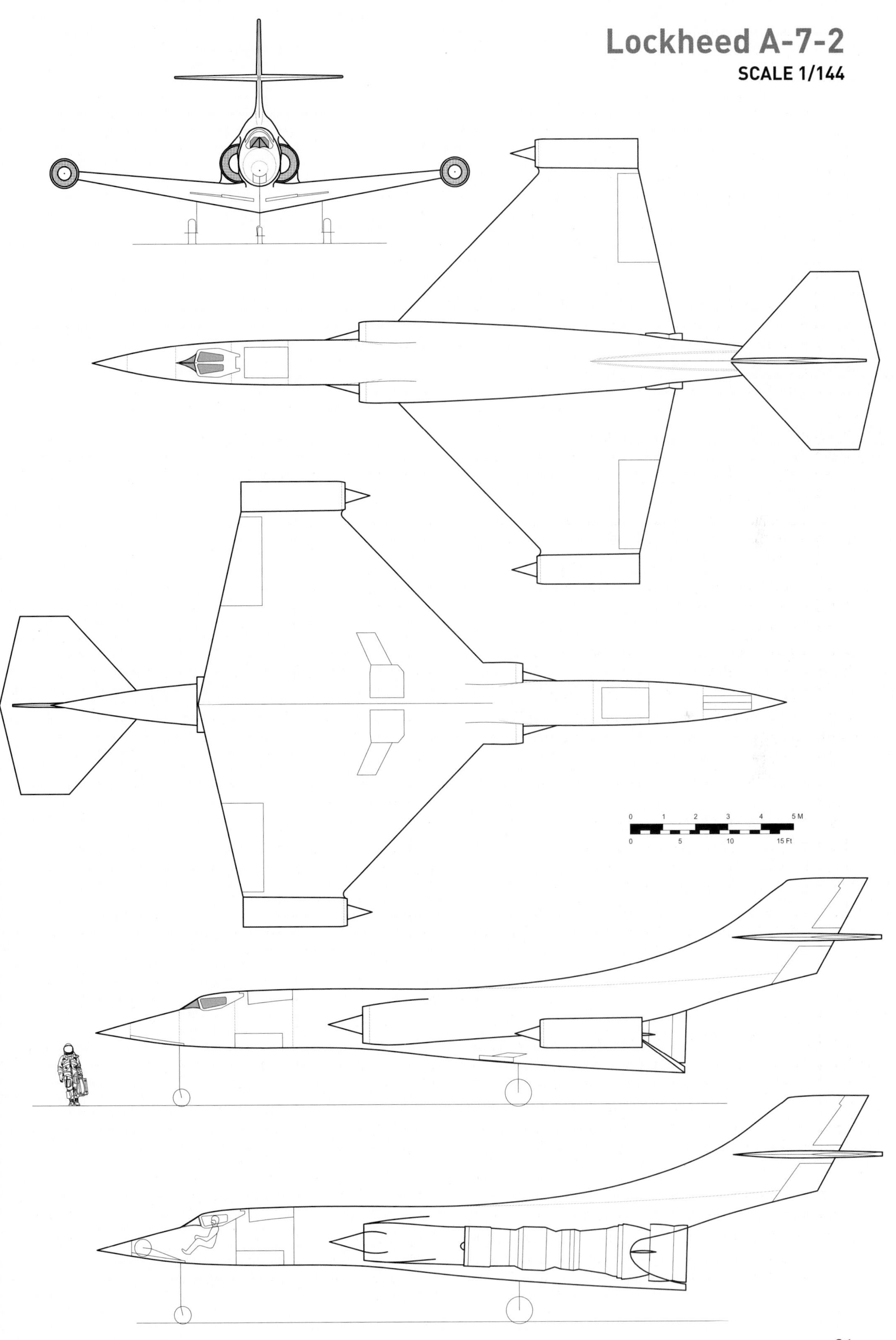
Lockheed A-7-2
SCALE 1/144
0 1 2 3 4 5 M
0 5 10 15 Ft

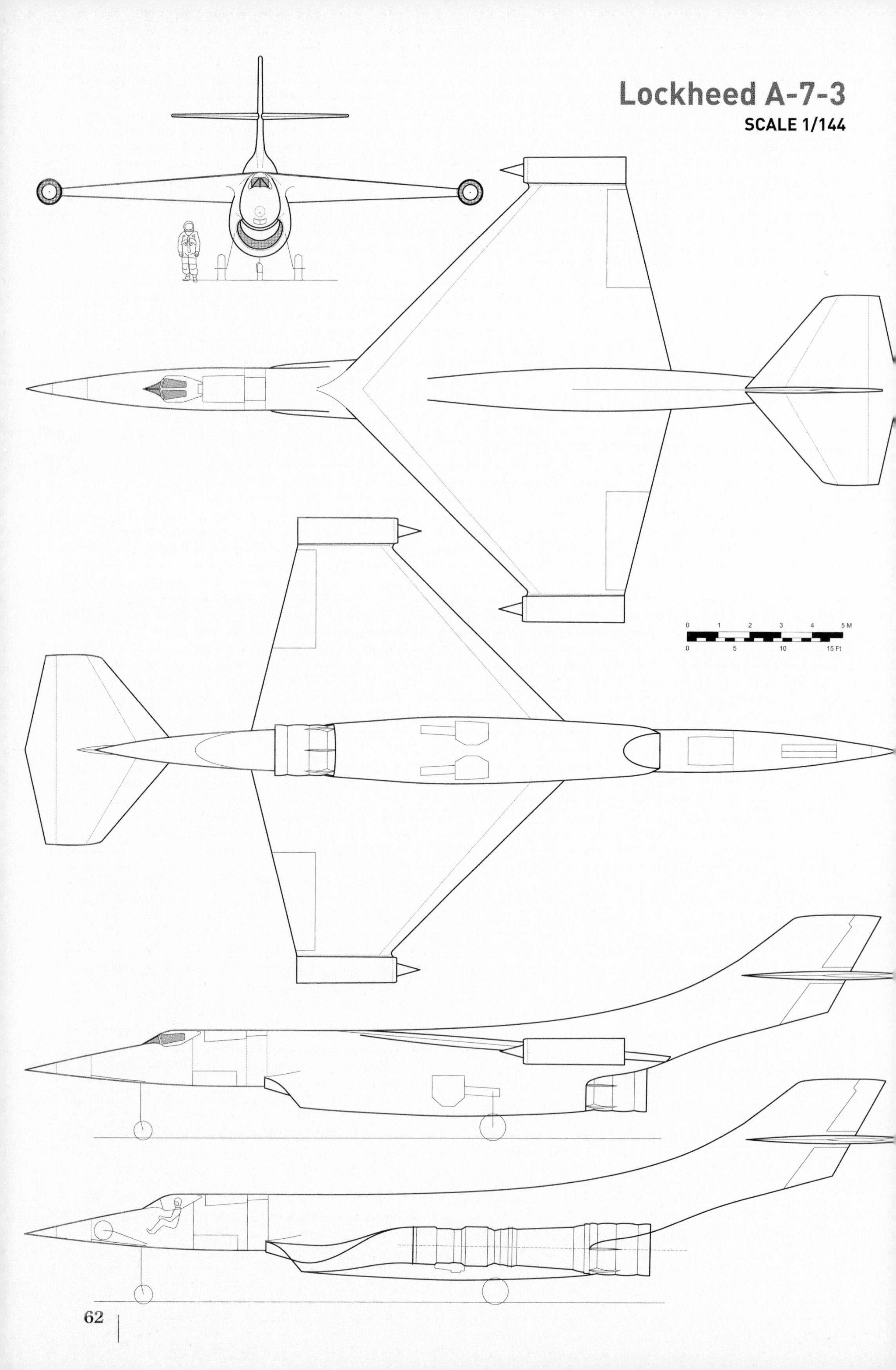
Lockheed A-7-3
SCALE 1/144
0 1 2 3 4 5 M
0 5 10 15 Ft

Lockheed A-10

SCALE 1/185

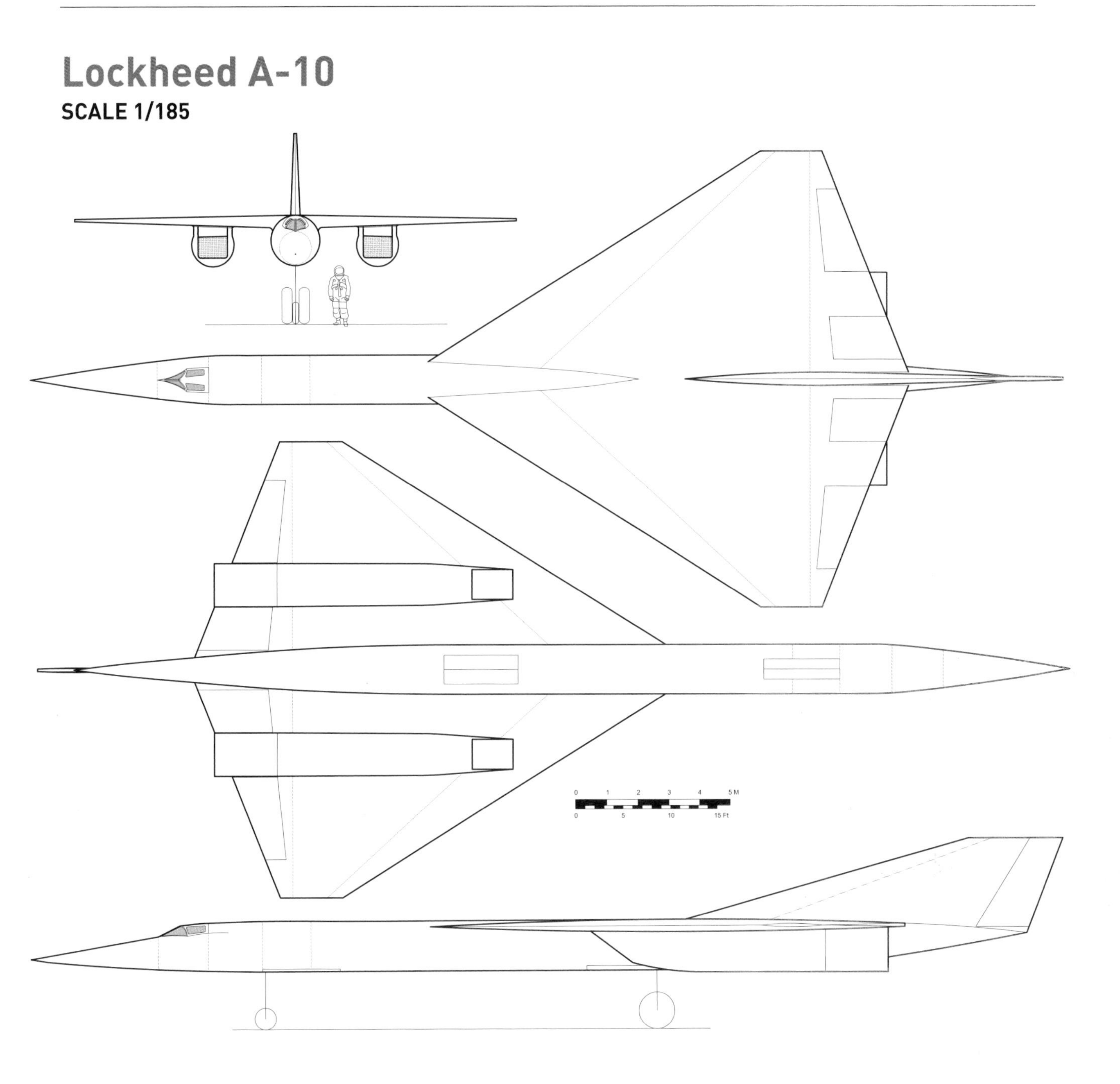

afterburning turbojets. However, even then the A-11 sought to do something unusual. In the baseline approach, the turbojets (both the J58 and J93 engines were studied) themselves would burn JP-150 jet fuel, but the afterburners – which would operate as ramjets once up to speed – would burn a high energy boron fuel. The fuel of choice would be HEF-3, ethlydecaborane ($C2H5B10H13$). As most of the flight would occur while under afterburner, most of the fuel carried would be HEF-3. The wings would carry 16,000lb of JP-150 and 2000lb of HEF-3, while the fuselage would carry 30,000lb of HEF-3.

A secondary approach would be to fill the tanks with only JP-150; this would not perform as well, but with a fuel load of 55,330lb and a gross takeoff weight of 92,130lb the A-11 could attain the range required. This would not require modifications to the aircraft apart from some materials. The baseline HEF-fuelled vehicles would have oversized tanks and would not normally fill them fully. If the tanks were filled to capacity with HEF-3 fuel, though, mission radius would extend to 2250 nautical miles.

The A-11 at last gave Lockheed a design that was considered to be definitively better than the staged concepts from Convair. An Air Force memo from June 5, 1959, compared the A-11 and the FISH point-by-point. Each plane had points of superiority, but on the whole the A-11 was considered superior. It would be cheaper to procure, cheaper to maintain, cheaper to operate. It would require fewer personnel to maintain and operate… it was estimated that the FISH system would require eight times the staff to fly a sortie. The recommendation was that FISH be abandoned in favour of the A-11.

While one of the recognized advantages of the A-11 over the FISH was the fact that it did not require a carrier aircraft, studies showed that using a second aircraft could substantially improve the performance of the A-11. By basing the A-11 at Edwards Air Force Base in California and meeting up with KC-135 tankers based in Fairbanks, Alaska, the A-11 could fulfill its recon missions without the need to operate from foreign bases. In order to accommodate inflight refuelling, the mission would be carried out using nothing but the JP-150 hydrocarbon fuel. Refuelling would occur between 25,000 and 40,000ft and between Mach 0.7 and Mach 0.82. With two refuellings – one before entry into Soviet territory, one following – virtually all of Russia could be surveyed. Exact range numbers are redacted from the available document, but given that the unrefuelled range of the A-11 was 4000 nautical miles, a version that is refuelled in flight twice, launching from southern California, flying a polar mission to, into and over Russia, turning around and flying back to southern California, would certainly have a far greater range.

Along with making foreign basing unnecessary, inflight refuelling would limit A-11 operations to a single base, simplifying not only operational concerns but also security operations.

A-12

The A-11 seemed to meet all the requirements except for radar cross section. It was in that aspect that the Convair FISH was clearly superior… and while FISH had been turned down, Convair soon came back with Kingfish. This would be a single-stage aircraft with the stealth properties of FISH, so Lockheed had a job to do – making the A-11 stealthy while retaining as much of its performance as possible.

Starting in early July 1959, Lockheed set to work on turning the A-11 into a stealthy aircraft. This new effort merited yet another designation change, this time to A-12. A diagram has been released by the CIA showing that someone had begun penciling in chines and a forward extension of the nacelle directly onto an A-11 diagram; this is the general direction taken by the A-12. However, the changes were more than cosmetic, and more than simply slapping radar absorbing and deflecting panels onto the A-11.

The A-11's wings were lowered to near, but not quite, the bottom of the fuselage; the engine nacelles were raised from beneath the wing to mid-wing, the 2D rectangular inlets replaced with circular ones that had large central translating spikes. The single vertical stabilizer was split in two, each new stabilizer mounted to an engine nacelle and tipped inboards to reduce radar reflection to ground receivers. And perhaps most importantly, the whole vehicle was blended as much as possible. The wings blended smoothly into the fuselage, the engines blended into the wings. As originally designed – and as the full scale mockup was first built – the forward fuselage was much like that of the A-10 and A-11, a largely featureless circular cross-section spike. But very quickly a change was made: sharp-edged chines ran all the way to the nose. These would serve both to reduce radar cross section (by installing radar absorbing wedges in the chines) and to improve aircraft lift/drag during cruise; they would also prove valuable locations for the installation of secondary instruments and sensors. The end result was an aircraft that looked like a cross between the high-performance A-11 and the low-RCS A-6.

The A-12 as designed was not quite the A-12 that would be built. As shown in the design created by Dick Fuller in roughly mid-July 1959, the vertical stabilizers were distinctly aft-swept; the fuselage tail extended quite far back. And the wingtips were, as with the A-11, sharply pointed, devoid of the conical camber and rounded planforms that would become famous. These features would be adjusted over the following weeks and months as the results of wind tunnel testing came in, but otherwise, the configuration was much like the A-12 that would actually be built.

The A-12 as designed in mid-July was able to cruise as fast as the A-11: Mach 3.2. But the cruising altitude and ceiling were notably lower, a tradeoff that was considered acceptable. The lower radar cross section would make the A-12 less visible than the A-11 even though it would be closer to radar transmitters and receivers. The A-12 would finally nail down the J58 as the engine of choice (with no apparent effort given to studying the J93), but with some new developments. Pratt & Whitney's engine had up until this point been more or less conventional, with a maximum speed of about Mach 2.5. But they designed a way to improve that. A 'bleed bypass' system, in the form of six external tubes, would duct about 20% of the compressor air aft of the fourth stage past the turbine section, straight to the afterburner. The effect would be to cool the actual metallic structure of the afterburner, allowing the combustion process to be run much hotter. This improved afterburner thrust and specific fuel consumption without increasing engine frontal area and with only a minor increase in weight.

The performance, in terms or range, speed, altitude and RCS of the A-12 was considered good enough compared to the Convair Kingfish, with more evidentiary basis. In late August of 1959 the Air Force awarded Lockheed a contract to go ahead with development of the A-12. At last the Mach 3 airplane was about to be born.

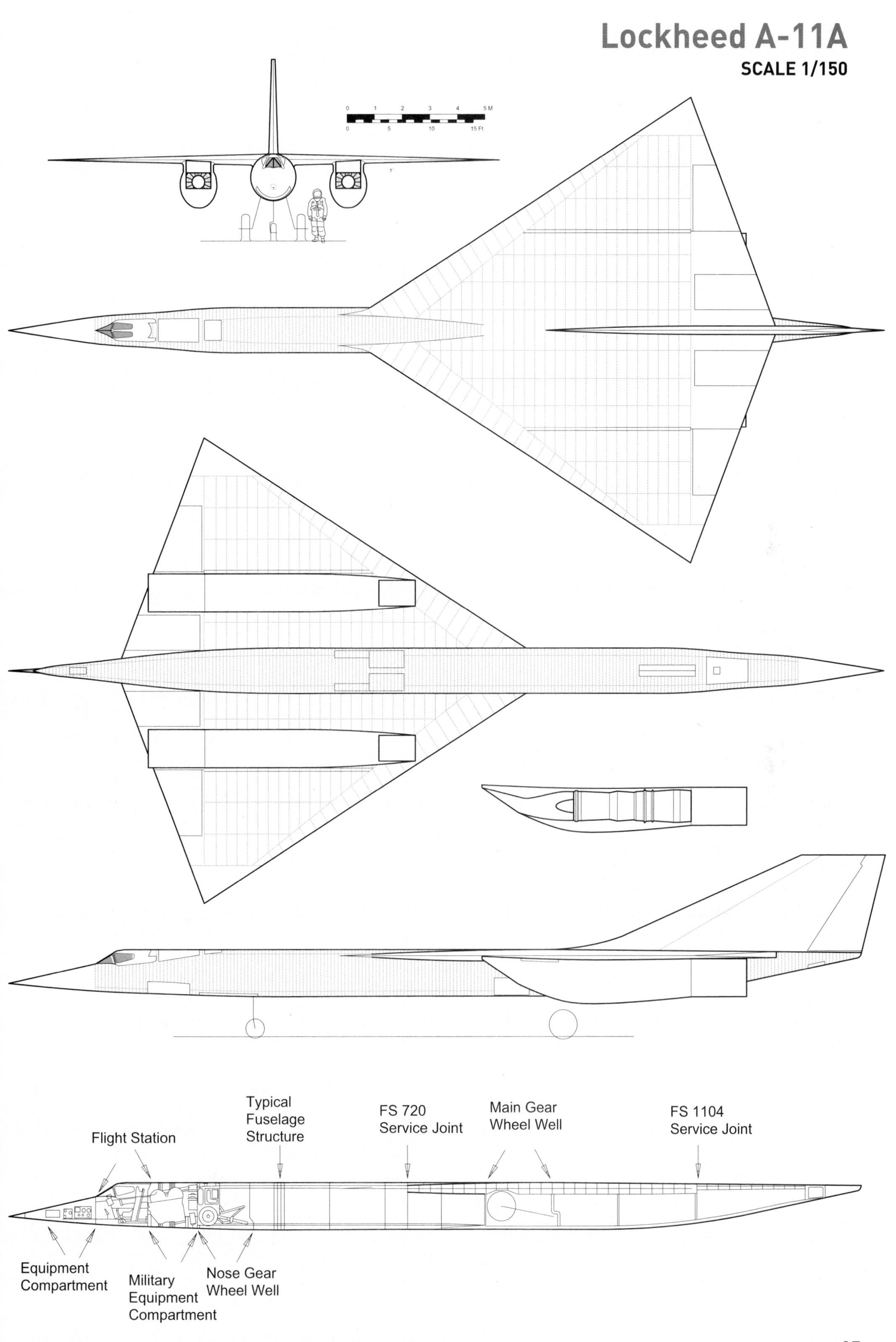
Lockheed A-11A
SCALE 1/150
0 1 2 3 4 5 M
0 5 10 15 Ft
Typical
Fuselage
Structure
FS 720
Service Joint
Main Gear
Wheel Well
FS 1104
Service Joint
Flight Station
Equipment
Compartment
Military
Equipment
Compartment
Nose Gear
Wheel Well

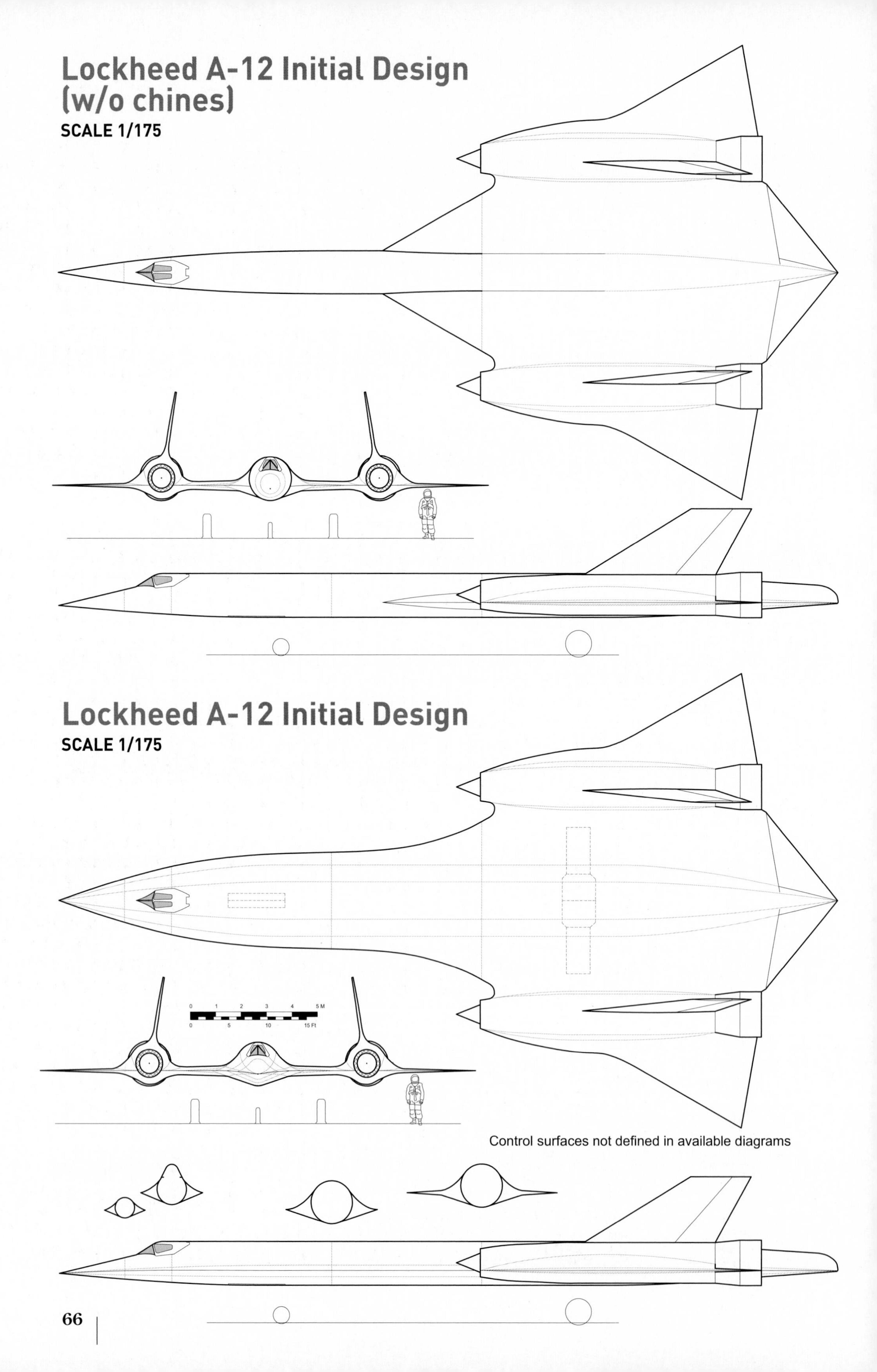
Lockheed A-12 Initial Design (w/o chines)
SCALE 1/175
Lockheed A-12 Initial Design
SCALE 1/175
0 1 2 3 4 5 M
0 5 10 15 Ft
Control surfaces not defined in available diagrams

4 The Concept Takes Flight

Once the A-12 design was selected to proceed to production, the real work of detailed design refinement began. This was specified in February of 1960 with a CIA contract to build a dozen A-12 reconnaissance airplanes with one being a two-seat trainer. The A-12 concept was given a full series of wind tunnel and radar cross section model tests, with the result that the design evolved somewhat but not enough to change the A-12 designation. The trailing edges of the swept-aft vertical stabilizers of the initial A-12 concept were instead swept forward; this allowed the wings to shield the entirety of the vertical stabilizers from ground based radar. The separate rudders were deleted from the vertical stabilizers; instead, the stabilizers were made all-moving, making them much more efficient in the thin high altitude air. The boat-tail between the engine nacelles was cut shorter in order to deal with a stability issue produced by the chines.

Evidently the difficulty with stability was at one point in the design process severe enough to merit the wind tunnel testing of a major configuration departure. A photo depicts a steel wind tunnel model of the A-12 with sizable highly-swept canards mounted just aft of the cockpit. The rest of the configuration appears to be halfway between the original A-12 concept and the final design, with the more curved planform just inboard of the engine nacelles and a lack of conical camber on the outboard wing panels.

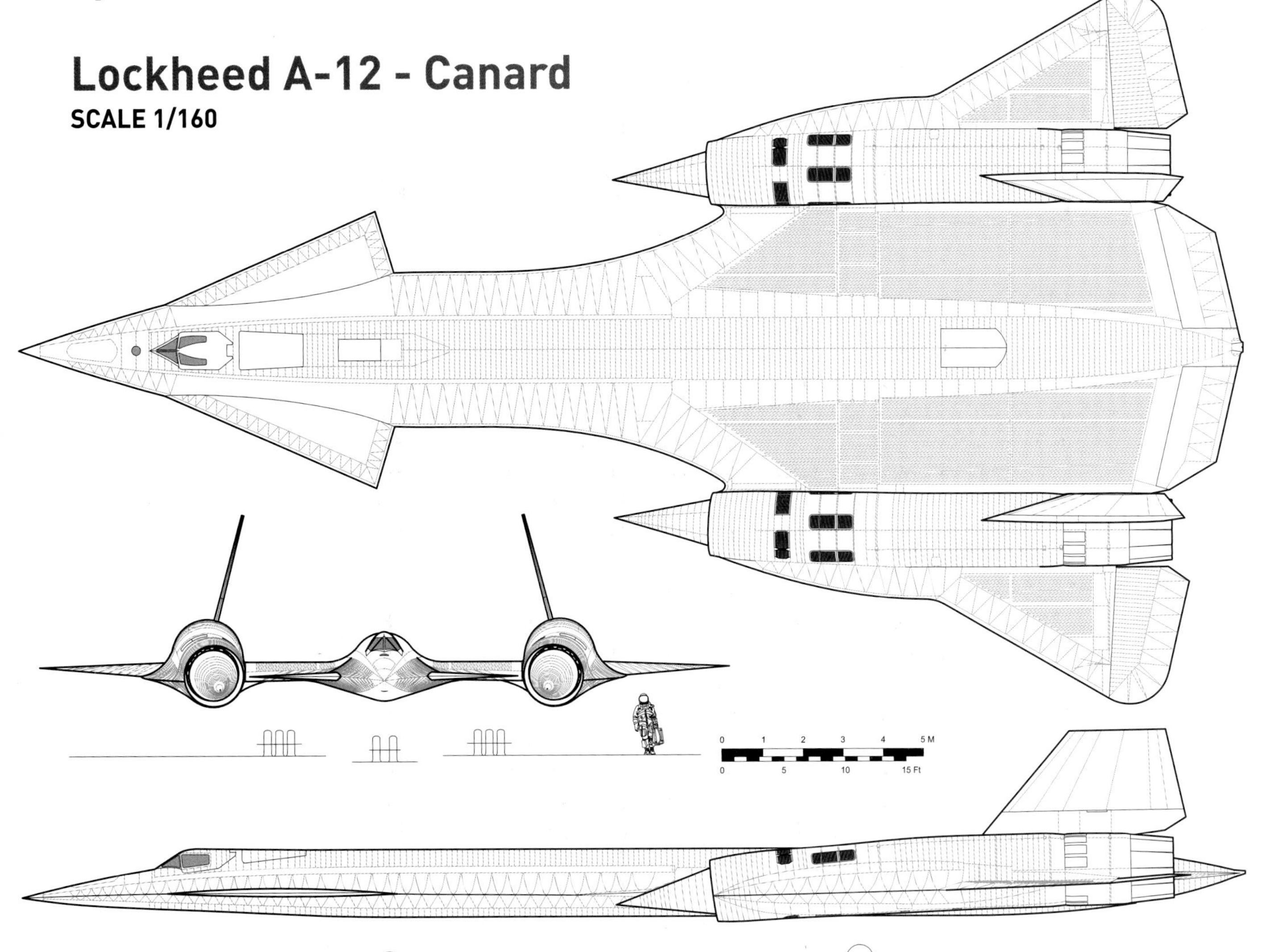

The need for the A-12 was driven home quite soon. In May 1960, a U-2 piloted by Francis Gary Powers was shot down over the Soviet Union. This was to have been a ground-breaking mission for the U-2, flying further than ever before, deeper into Soviet territory and in fact all the way across the country. It was a risky mission but considered to be worth the risk for the trove of photographic intelligence it would gather. Unfortunately, the new Soviet SA-2 Guideline surface-to-air missile proved to be up to the task; three were fired at the U-2 as it flew over the Urals, the first fatally crippling the craft. The wreckage was recovered, as was the pilot.

It was a propaganda coup for the Soviets: until that point, they had known that the U-2 was flying over their territory but they could not prove it. With a pilot, wreckage, advanced surveillance cameras and exposed film on hand, they could – and did – go before the world and raise hell. The U-2, as good a platform as it was, had proven that it could not fly high enough to be considered invulnerable. Something less visible flying higher and faster was needed, and the A-12 was already in development to be just the U-2 replacement that the CIA and the USAF wanted.

Production of the A-12 began at the Lockheed plant B-6 in Burbank, California, in Building 309/310. A search for pilots began – a process of vetting for both proficiency and security issues. And work began at Area 51 to build new hangars and other facilities, as well as a 12,000ft runway extension.

The story about how the CIA used devious schemes to procure the titanium the programme needed from the Soviets has been often and adequately told, so it doesn't bear repeating here. Suffice it to say, about 93% of the aircraft was fabricated from titanium. In the 1950s (and in the decades since) titanium was seen as an almost magical material, with the strength of steel and the density of aluminum, with the benefit – depending on application – of a very high melting temperature. But in order to properly achieve these properties, the right alloy must be used, as pure elemental titanium is not only substantially weaker, it also has the unfortunate feature of being flammable.

Chips of titanium, an inevitable result of common manufacturing processes, are a substantial fire hazard... not only catching fire easily, but burning with a white heat that can strip the oxygen out of water molecules, making extinguishing the fire problematic. As such, more easily machinable and less flammable alloys were quickly produced. As referenced in the Archangel chapter, B120VCA (also known as Ti-13V-11Cr-3Al) was an early alloy for aerospace applications. It is difficult to work, with poor weldability, but it makes for great skins and springs and is both cold formable (i.e. it is ductile and can be shaped and trimmed at low temperatures) and heat treatable. It is this latter feature that proved perhaps most interesting: with each high speed flight of an A-12 or one of its derivative aircraft, the aerothermal heating would end up heat-treating the structure, restoring its strength rather than weakening it.

At this early stage, much about the practical working or titanium was unknown; Lockheed's engineers had to learn for themselves that most commercial cutting fluids sped up stress corrosion on hot titanium and that the chlorine in Burbank's water, intended to cut down on algae growth, also cut down on successful spot welding of titanium parts. Felt-tipped Pentel pens had chlorine in the ink which caused titanium parts to be etched if marked. Cadmium coated wrenches left behind a nearly microscopic trace of cadmium on titanium bolts... enough to cause the bolt heads to drop off when heated. It was found that virtually everything the parts might come into contact with, from pens and pencils, masking tape, plastic, paint and so on, could all contain contaminants that would cause titanium embrittlement and structural failure.

Determining the radar cross section of the A-12 involved the use of models, ranging from 1/8 scale up to full size. The full size RCS model was constructed of wood covered with sheet metal; the end result being a passably realistic mockup. Testing of the models helped refine the overall aircraft configuration as well as the detailed design elements used to reduce the signature of the aircraft.

The A-12 had two distinctly different approaches to reducing radar cross section. The leading and trailing edges of the chines and wings were initially fitted with the same sort of triangular 'teeth' that characterized the Convair FISH and Kingfish. Filling the space between the teeth would be a radar absorbent foam, structurally adequate to withstand low speed dynamic pressures, but not high speed. This would allow the A-12 to get into the air and begin flight testing and radar testing. For higher speed flight the foam would be replaced with sheet metal triangles... this would not be stealthy at all, but would serve structurally throughout the performance envelope.

At a later time, the teeth along the chines would be replaced with radar absorbent rectangular composite honeycomb panels, good for operational use. The initial plan was that these would be fibreglass impregnated with graphite, but after considerable testing the panels were made from a composite of asbestos and silicone. The outer skin was a 3/32in thick baked asbestos mat, backed up by an asbestos/silicone honeycomb infused with nearly microscopic graphite spheres. The sawtooth foam inserts in the leading and trailing edges of the wings were replaced with an improved plastic that absorbed radar while also surviving the flight conditions of the craft.

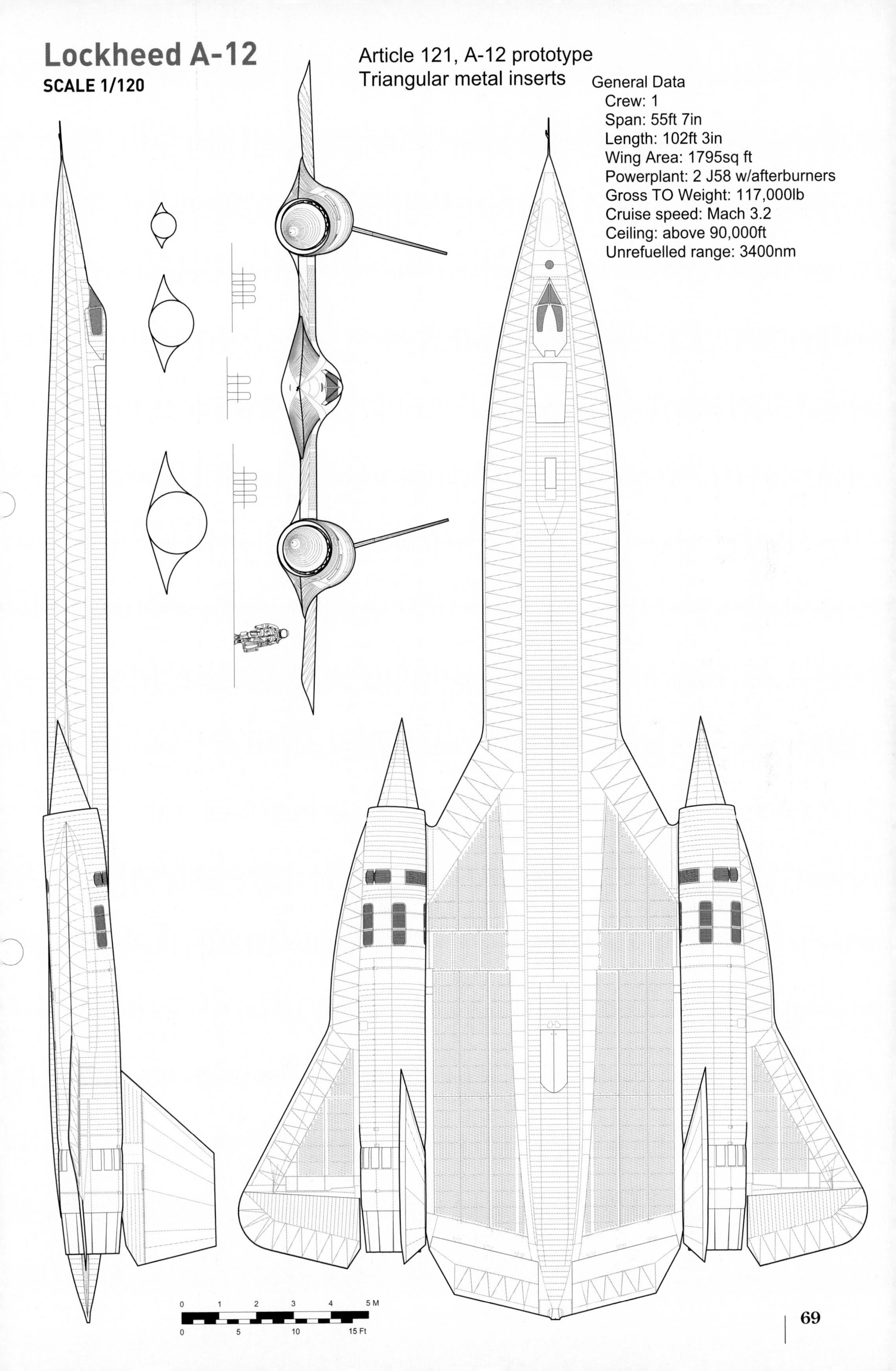
Lockheed A-12
SCALE 1/120
Article 121, A-12 prototype
Triangular metal inserts
General Data
Crew: 1
Span: 55ft 7in
Length: 102ft 3in
Wing Area: 1795sq ft
Powerplant: 2 J58 w/afterburners
Gross TO Weight: 117,000lb
Cruise speed: Mach 3.2
Ceiling: above 90,000ft
Unrefuelled range: 3400nm
0 1 2 3 4 5 M
0 5 10 15 Ft

Lockheed A-12
SCALE 1/120

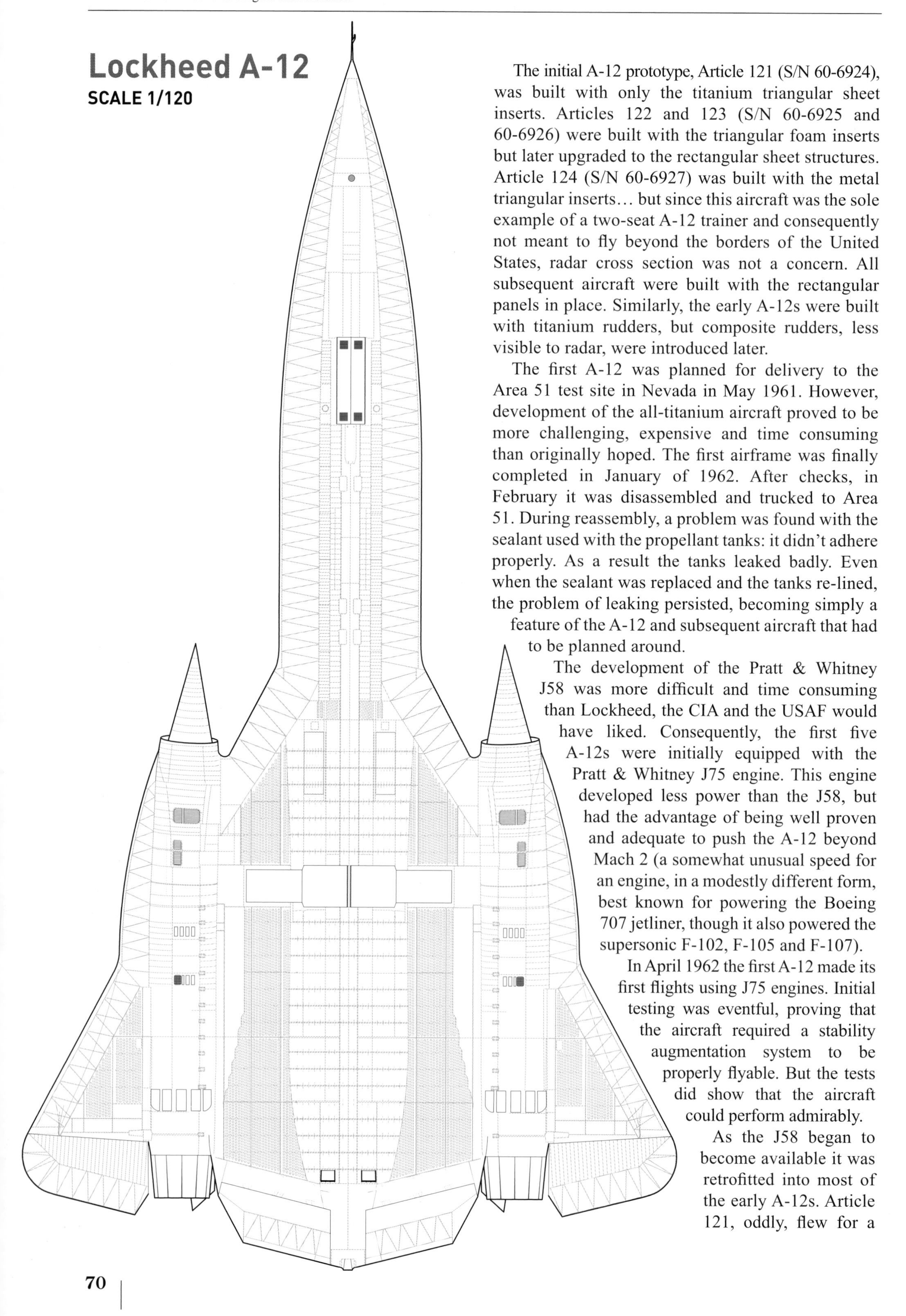

The initial A-12 prototype, Article 121 (S/N 60-6924), was built with only the titanium triangular sheet inserts. Articles 122 and 123 (S/N 60-6925 and 60-6926) were built with the triangular foam inserts but later upgraded to the rectangular sheet structures. Article 124 (S/N 60-6927) was built with the metal triangular inserts… but since this aircraft was the sole example of a two-seat A-12 trainer and consequently not meant to fly beyond the borders of the United States, radar cross section was not a concern. All subsequent aircraft were built with the rectangular panels in place. Similarly, the early A-12s were built with titanium rudders, but composite rudders, less visible to radar, were introduced later.

The first A-12 was planned for delivery to the Area 51 test site in Nevada in May 1961. However, development of the all-titanium aircraft proved to be more challenging, expensive and time consuming than originally hoped. The first airframe was finally completed in January of 1962. After checks, in February it was disassembled and trucked to Area 51. During reassembly, a problem was found with the sealant used with the propellant tanks: it didn't adhere properly. As a result the tanks leaked badly. Even when the sealant was replaced and the tanks re-lined, the problem of leaking persisted, becoming simply a feature of the A-12 and subsequent aircraft that had to be planned around.

The development of the Pratt & Whitney J58 was more difficult and time consuming than Lockheed, the CIA and the USAF would have liked. Consequently, the first five A-12s were initially equipped with the Pratt & Whitney J75 engine. This engine developed less power than the J58, but had the advantage of being well proven and adequate to push the A-12 beyond Mach 2 (a somewhat unusual speed for an engine, in a modestly different form, best known for powering the Boeing 707 jetliner, though it also powered the supersonic F-102, F-105 and F-107).

In April 1962 the first A-12 made its first flights using J75 engines. Initial testing was eventful, proving that the aircraft required a stability augmentation system to be properly flyable. But the tests did show that the aircraft could perform admirably.

As the J58 began to become available it was retrofitted into most of the early A-12s. Article 121, oddly, flew for a time with one J75 and one J58 starting in October 1962. Article 124 never replaced the J75s, instead flying training missions throughout its life with the less powerful engines. Article 121 first flew with a pair of J58s in January, 1963.

As more J58s and A-12 airframes became available, the test flight programme expanded. And while the design proved itself capable, it also proved itself troublesome with numerous lost aircraft. Article 123 was lost in May of 1964; Article 133 in July of that year; Article 126 in December. The losses were not due to inherent flaws in the design, but to issues as mundane as ice in the pitot system and incorrect wiring. But despite these troubles the A-12 continued to hit important milestones. In May 1962 an A-12 (Article 121) went supersonic for the first time; in July 1963 Mach 3 was attained. In November 1963 it reached design speed and altitude, and in February 1964 it cruised at Mach 3.2 and 83,000ft for 10 minutes. And at last in January 1965 an A-12 cruised at more than Mach 3.1 for an hour and 15 minutes.

The J58, Pratt & Whitney internal designation JT11D-20A, was an impressive engine. But it used a unique fuel, JP-7, which was composed (primarily of paraffins and cycloparaffins) to have a low vapour pressure and a high flash point in order to be stable in the extreme conditions in which the aircraft would operate. The fuel was expected to get as hot as 350°F due to aerothermal heating of the airframe, which would cause most jet fuels to boil, decompose or even explode. That is not the case with ultra-stable JP-7.

Famously, pails of this fuel would not ignite if lit matches were thrown into them due to the low vapour pressure. This made the fuel safe both on the ground and in flight, but it also made it irritatingly difficult to light. So to assure ignition an auxiliary chemical ignition system included a small canister (600cc/1¼ pint) of triethylborane (TEB, *C6H15B*) in each nacelle. TEB is a pyrophoric chemical, meaning it will spontaneously combust when exposed to air above 0°F. The TEB enthusiastically burns with a green flame, with a combustion hot enough – around 3000°F – to ensure that the reluctant JP-7 would ignite. TEB is a dangerous substance to work with; not only will it spontaneously combust when in contact with air, it is also toxic, causing tissue damage on contact. Fighting a TEB fire is difficult because it will also react with many fire-fighting chemicals such as carbon tetrachloride and halogens. Fortunately, though, it does not react with water and water-based foams. However, TEB is substantially less dense than water so it will float to the surface, be exposed to air, and burst into flames again.

In the A-12, a small metered quantity of TEB would be nitrogen pressure-injected into either the main burner (on engine startup) or the afterburner. Enough TEB was included for 16 shots, which were automatically fired based on throttle position. TEB, generally mixed with triethylaluminum, has been used in a similar capacity with a number of liquid propellant rockets, such as the Saturn Vs F-1 and the SpaceX Falcon 9 Merlin.

Just getting the J58s to start was a non-trivial undertaking. Spinning up the turbines was not a job for a small built-in starter motor; instead a pair of race car motors, totalling more than 600hp, were used.

Much of the success of the J58 came from the structure around it. Simply hanging the engine out in the breeze would not have extracted optimum performance, as the turbine could not operate with a supersonic airflow slamming into it. Thus the engine was located in a nacelle with an inlet spike. At supersonic speeds the spike would shed a shock wave from the tip and the air flow relative to the engine would be slowed to subsonic speeds by the shock wave.

The inlet spikes on early A-12 test flights were fixed in place, but as the development programme progressed the spikes were able to translate forward and aft. This was to ensure that the conical shockwave coming from the tip of the spike was 'swallowed' by the circular inlet. The inlet spike begins moving at Mach 1.6, translating aft 1 5/8in for each 0.1 Mach, for a total translation of 26in at top speed. The inlet spikes were not pointed directly forward. Instead they were canted inboard by 3° to account for the effects of airflow coming from the forward fuselage, and pointed down 5° due to the angle of attack the aircraft would cruise at.

The nacelle included a number of doors permitting air to flow in and out. Above and below the inlet were a ring of bypass doors which would be used to reduce excess air pressure in the inlet ahead of the turbine. Further back atop the nacelles were inlets that fed a duct that ran into the inlet spike, and out again through a series of small slots at the widest part of the spike.

At the extreme rear of the nacelle was a ring of adjustable 'turkey feathers'. On standard military aircraft, these form the exhaust nozzle for the afterburner and are part of the engine. On the A-12, these were actually part of the airframe, and were aft of the afterburner. The turbojet and afterburner were an integrated unit with its own adjustable exhaust nozzle, but aft of that was an ejector section used to increase thrust further at zero fuel cost.

Just ahead of the ejector section was a ring of suck-in doors, serving as secondary inlets. Starting at about Mach 1.2, air would be brought in and mixed with the exhaust from the afterburner, increasing mass flow and thrust. The J58 engine was suspended within the nacelle, in the middle of an air duct that passed around the engine. Air would flow around the engine brought in from shock traps just aft of the inlet throat, providing some cooling action for the engine and air flow to the ejector.

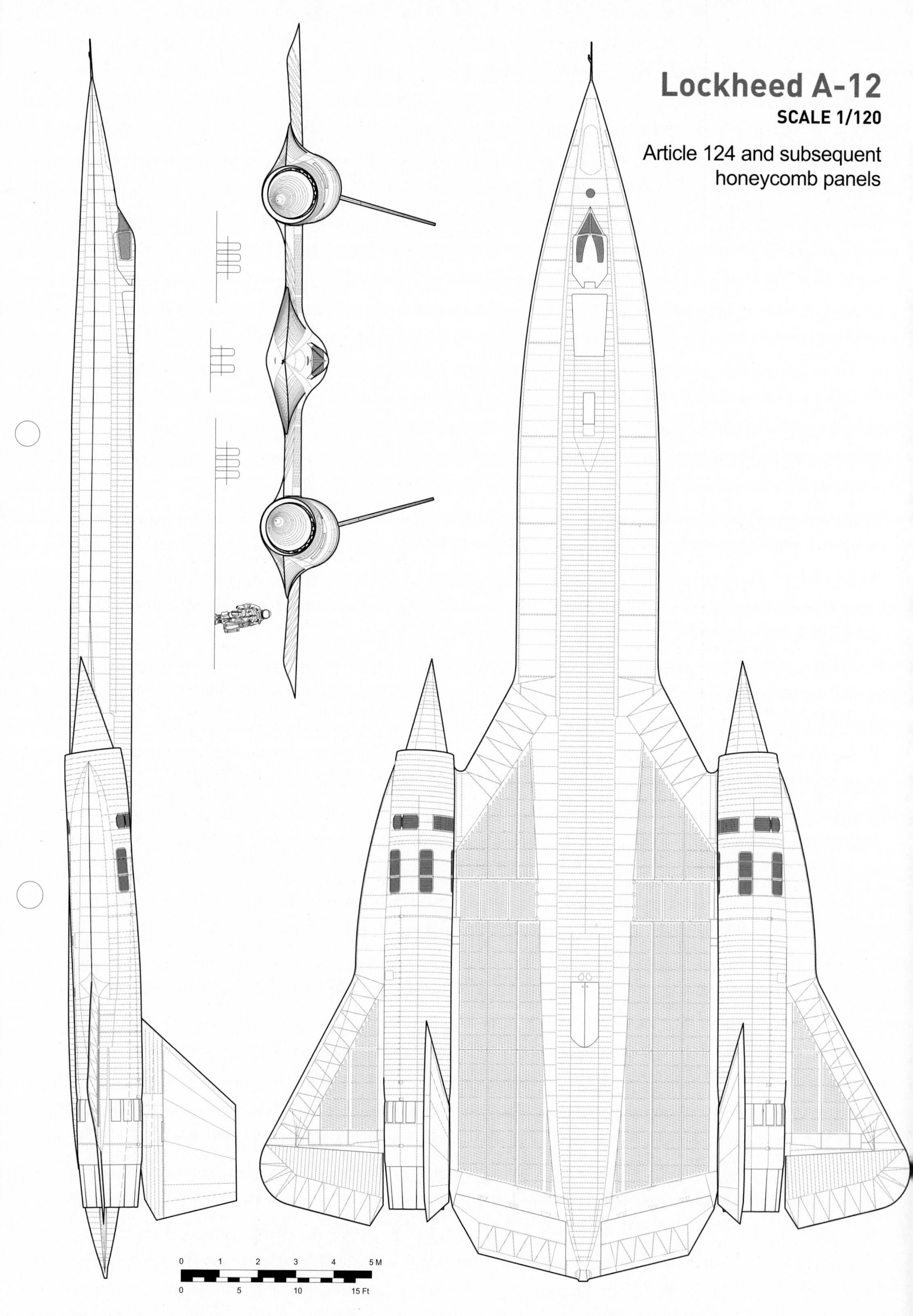
Lockheed A-12
SCALE 1/120
Article 124 and subsequent honeycomb panels
0 1 2 3 4 5 M
0 5 10 15 Ft

The A-12 was a thirsty aircraft, requiring substantial fuel storage. Most of the fuselage was taken up with six fuel tanks, the first behind the air conditioning equipment just aft of the Q-bay, the last almost to the tip of the tail. Tanks 4, 5 and 6 had extensions that ran into the wings. An in-flight refuelling receptacle was located on the upper surface of the fuselage just aft of the air conditioning equipment, feeding directly into the first tank. This receptacle was used not only in flight but on the ground for fuelling.

The engines had a schedule of which tank to draw from, using 16 boost pumps to supply fuel. The tanks were interconnected and the pumps could be used to shift propellant to maintain proper CG. If the fuel needed to be dumped, it would be ejected out through the extreme tail of the fuselage. Uniquely, the A-12 and its derivative aircraft used fuel as hydraulic fluid.

The A-12 was a world beater of a reconnaissance aircraft, capable of flying higher and faster than anything else in the world. To be a useful reconnaissance system, it needed to have useful reconnaissance systems. Directly behind the pilot's compartment was the 'Q-bay' containing the surveillance equipment and pressurized to 28,000ft. Four camera types with the film and film transport systems needed were developed, but only one, the Type I system, was actually used in operational missions. The Type I camera was a 690lb Perkin-Elmer panoramic stereo design, using two cameras simultaneously imaging onto a single piece of film. The lenses were 18in focal length f3.8 refractors, with pre-programmed exposures of 1/50 to 1/600 second.

The Type I system had a forward and an aft camera. The forward camera scanned the Earth directly below from 21° to the right of the nadir to 67° to the left of the nadir; the aft camera scanned the earth directly below from 67° to the right of the nadir to 21° to the left of the nadir. They peered out through quartz windows bonded to the aircraft structure.

The camera housing was lowered into the Q-bay through a hatch in the top of the fuselage and was kept at a constant temperature throughout the mission. The left and right views were not shot simultaneously but sequentially; there was, however, a fair amount of overlap in the coverage directly below the aircraft and between 21° left and right. The film was a single roll 6.6in wide by 5000ft long, sufficient for 1980 individual images and 2500 nautical miles of coverage.

Each image projected onto the film was a strip 27.67in by 6.36in, 1/60,000 scale at 80,000ft. At 80,000ft, the image would cover terrain 63 nautical miles wide with a resolution of 1ft at the nadir and 3ft at the far edge. The assembly was gimbal supported through the centre of gravity. The gimbal allowed the camera assembly to stabilize even if the A-12 was shaking and to remain pointed straight down even if the aircraft was climbing, descending or banking slightly. The supporting frame assembly was attached to the Q-bay structure.

In order to be an effective reconnaissance platform, the A-12 needed not only a good camera but the ability to get where it was meant to go. The primary navigation aid for the pilot was an inertial navigation system that relied upon three integrating gyros and three single-axis accelerometers mounted to a gyro-stabilized platform to calculate the position of the aircraft based upon direction and accelerations. Position and ground speed were displayed along with distance and direction to up to 42 pre-selected targets. A backup flight reference system used a single axis directional gyro stabilized by a two-axis vertical gyro, slaved to a compass transmitter which detected the Earth's geomagnetic field.

Regardless of how well designed and manufactured it might be, an inertial guidance system inevitably introduces cumulative errors. The more manoeuvring, the more vibration, the more errors. But in the era before GPS, this was about the best that could be hoped for. So in order to perfect and correct the navigation capability of the aircraft, the pilot had a periscope looking through a port in the underside of the nose. This provided a view straight down and somewhat forward with two settings: a wide field view providing a view 85° forward of the nadir (at 90,000ft altitude, looking 242 nautical miles forward, 32 nautical miles laterally) and a narrow field view 47° forward of the nadir (17 nautical miles forward, 7.9 nautical miles laterally). This was to aid in navigation via update fixing the inertial navigation system.

The periscope would be used to spot preselected ground objects; comparisons would be available with a 35mm film strip map which could be projected onto the same 6in display. The display could be toggled to look upwards through a port just ahead of the canopy. This mode would be used in conjunction with a sun compass to provide directional data based on a pre-computed position of the sun.

If the aircraft went down over enemy territory, the film strips could be destroyed with a thermite charge that would burn at 2000°F. The same self-destruct command would use pressurized nitrogen to force water into the map case; the water soluble paper maps would be turned to slush. This would prevent enemies from knowing what the US knew about their territory and what was of interest to the CIA and USAF.

Even though the A-12 was designed to be stealthy, high and fast, it was accepted that the Soviets would eventually end up spotting the plane and taking pot-shots at it. The most likely system for the Soviets to use was the SA-2 Guideline surface-to-air missile, introduced in 1957. The SA-2 was a two-stage missile designed specifically to knock down high-altitude targets up to 82,000ft. The missile was controlled by

Perkin Elmer Type I Camera

SCALE 1/30

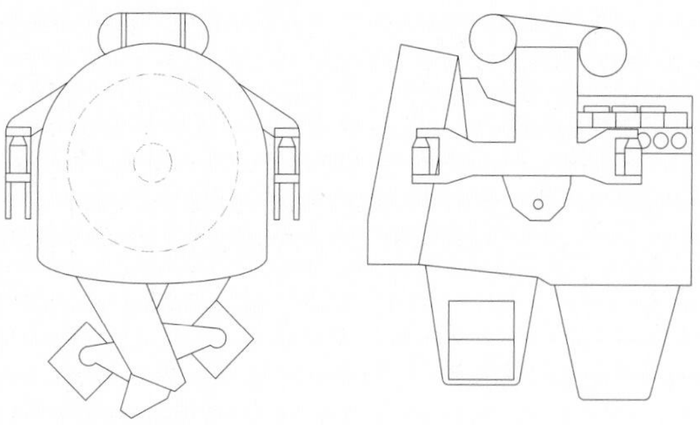

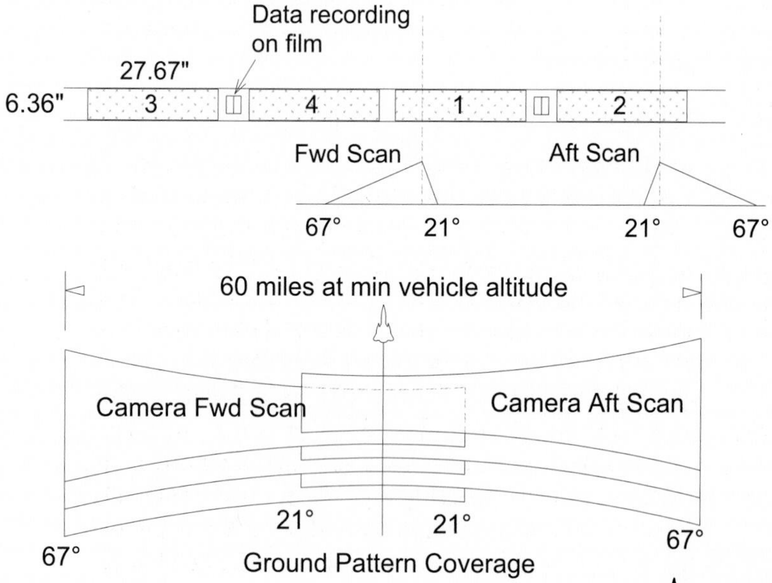

Data recording
on film
27.67"
6.36"
3
4
1
2
Fwd Scan
Aft Scan
67°
21°
21°
67°

60 miles at min vehicle altitude
Camera Fwd Scan
Camera Aft Scan
21°
21°
67°
67°
Ground Pattern Coverage

Lockheed A-12

SCALE 1/120

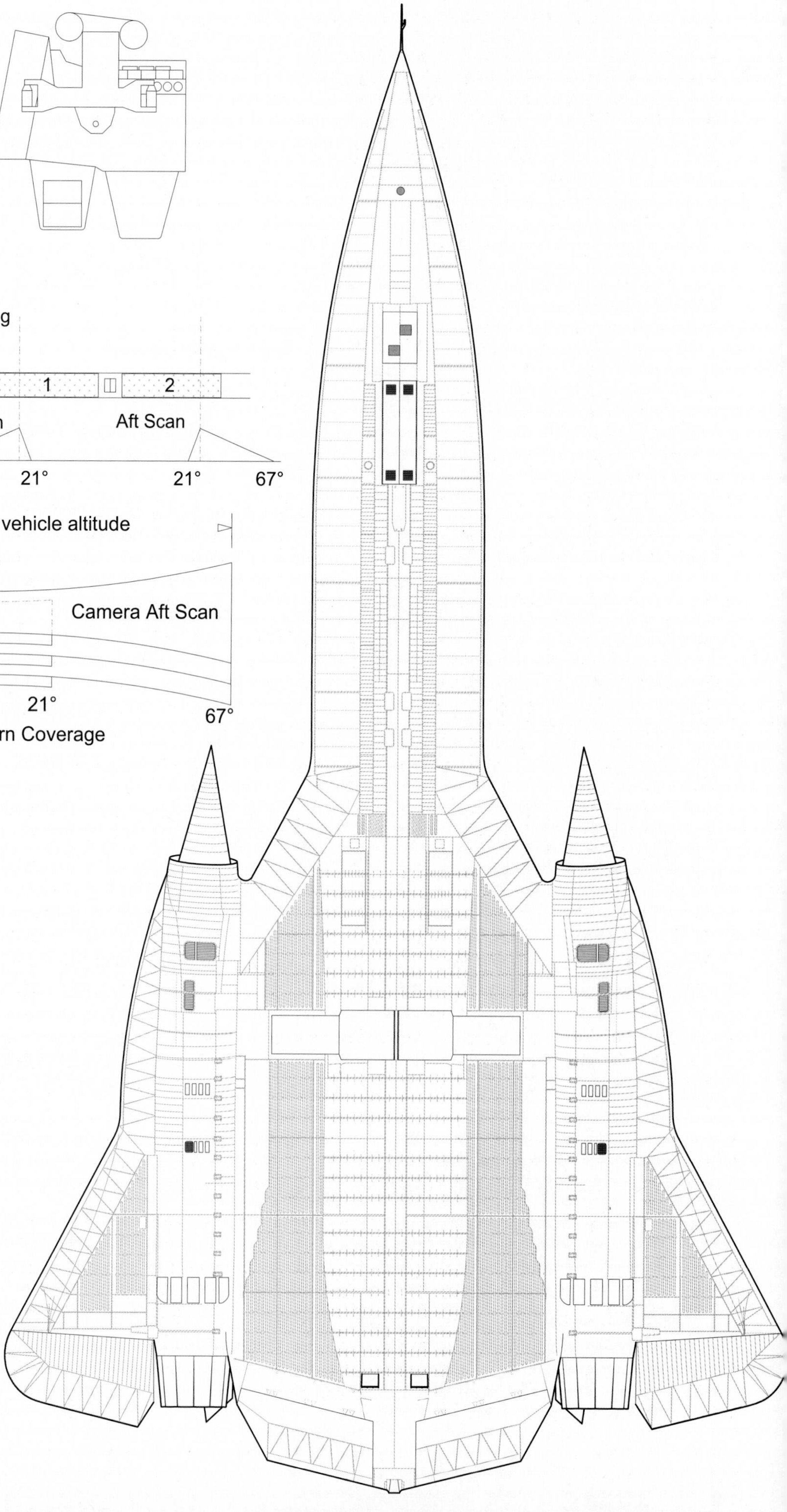

a command transmitter on the ground which radioed guidance orders to it. The United States had encountered it before: when one shot down Francis Gary Powers' U-2 in 1960 and another U-2 over Cuba in October 1962. Consequently the A-12 had an electronic countermeasures suite tailored specifically to defeat the SA-2. The 'Supermarket' ECM suite included:

- Pin Peg: A radar receiver system that provided position data to the pilot of SAM tacking radar station. A 30lb system built by Westinghouse, it could receive in the range of 2.8 GHz to 3.2 GHz and 4.8 GHz to 5.2 GHz.
- Blue Dog: This would record the L-band command signals sent to the SA-2 and play them back, confusing the missile during terminal guidance. It was a 480lb instrument built by Sylvania with an output of 20 kilowatts.
- Big Blast: An SA-2 barrage jammer that would send out high-power 'noise' in the range of 2.8 GHz to 3.2 GHz and 4.8 GHz to 5.2 GHz. This would be a last-resort system since it would make the aircraft blindingly obvious. Weighing 400lb, it was built by Applied Technology, Inc. and was activated automatically by Pin Peg or Blue Dog.
- Mad Moth: An ill-defined (in available literature) deceptive jammer, used as an alternative to Big Blast.

The ECM systems were located in compartments in the chines outboard of the fuselage. Big Blast and Mad Moth would go in the forward-port compartment; Blue Dog would go in the forward starboard compartment. There were a number of other instruments with unenlightening code names, including Fat Fox (forward starboard compartment); Bass Bug (forward port); Occasion, Sam Spade, System 17, and System 21 (aft starboard); Ice Bag (on the lower Q-Bay hatch); and Cat Nap (in the chines just forward of the Station 715 field joint on both sides). Other than a diagram showing where these items were, there is little to describe them, though Sam Spade had both transmitters and receivers, while Cat Nap appeared to possibly have openings out the side, indicating either side-looking optics or air sampling systems.

The pilots were given pressure suits that were not far from being early space suits. Earlier flight suits worn by US pilots of airbreathing aircraft had been at most partial pressure suits, but the A-12 would fly in a whole new regime: not only at extreme altitude but also at extreme temperatures. The canopy would be exposed to extreme aerothermal heating and was expected to reach oven-like temperatures; this would make the cockpit extremely hot, even with air conditioning. And if the pilot had to eject at altitude, he would be suddenly exposed to shockingly low temperatures.

Also, in order to keep weight down, the cockpit would depressurize as the aircraft gained altitude, finally stabilizing internal pressure at the equivalent of 26,000ft (5.22 psia), just 3000ft lower than the top of Mt. Everest. So the David Clarke Company was contracted by the CIA to provide an all-new full pressure suit (5.68 psia), the S-901. This initially bore a great deal of resemblance to the space suits worn by X-15 pilots and Mercury astronauts, as it had a modern 'space suit' helmet and a reflective aluminized outer layer of Nomex (to help it stay cool in the oven-like cockpit). The S-901 was produced in 12 standard sizes, rather than being tailor-made like the space suits astronauts wore. The exterior covering would eventually be replaced with a white Dacron outer garment to eliminate bothersome reflections onto the instrument panel.

The pilot was equipped with a rocket-powered ejection seat for emergencies. One would be forgiven for thinking that wind blast at Mach 3+ would be a major issue; after all, the pilots of such Mach 3 aircraft as the B-70 and the proposed Republic XF-103 and North American F-108 were to be equipped with ejection capsules, as were the crew of the Mach 2+ Convair B-58. But at around 100,000ft, the air is so thin that the dynamic pressure at Mach 3 is the equivalent of only 175 knots at sea level, easily survivable for a man in a good pressure suit. The ejection seat was good from sea level to maximum latitude, but in order to ensure proper parachute inflation and a safe landing speed, ejection while on the ground would have to occur at more than 65 knots.

In case the pilot needed to check on the condition of the vertical stabilizer or the engines, a rear view periscope could be manually extended. The periscope, normally locked into a retracted position in the top of the canopy, would provide a 10° cone of visibility, with the ability to rotate left or right by a further 10°.

The A-12s went through a series of changes in their paint scheme. As originally built, they were unpainted, their titanium structures on full display. For low speed test flight this was perfectly adequate. But at higher speeds heat would start to become an issue, in particular for the structures made out of composite materials. So these areas – the chines, inlet spikes, wing leading and trailing edges, the rudders for those vehicles with composite rather than titanium rudders – were coated with a blue-black paint with an emissivity of 0.93. Black would radiate heat away, helping to keep the plastic structures cool. Eventually the decision was made to coat the entire structure in the black paint, helping to cool the titanium structure as well. This of course led to the most famous descendent of the A-12 being called the 'Blackbird', though the A-12 itself does not seem to have been called that. The windshield was coated with low-reflectivity magnesium fluoride.

For all the effort and resources poured into the A-12, its service life was remarkably stunted. It was designed to overfly the Soviet Union, but it never did. Through 1965 and 1966 proposals were made to operationally overfly China and South East Asia, but technical and political concerns stymied efforts to use the A-12. By early 1967, plans were already starting to come together to mothball the whole fleet. But somewhat out of the blue concerns about surface-to-air missile deployment in North Vietnam raised hopes of the A-12 finally being used. And in May 1967, deployment began to Kadena air Force Base in Okinawa, Japan. The first operational mission was carried out on May 31, 1967, cruising at 80,000ft and Mach 3.1 over North Vietnam.

Numerous supposed SAM sites were photographed and it appeared that enemy radar systems had not detected the aircraft. Later flights were detected; in December of 1967 an A-12 was fired upon by six surface-to-air missiles, with the result that a piece of threaded rod from one of the detonating missiles embedding itself in a lower wing root. The last operational A-12 flight occurred on May 8, 1968, and the very last flight of any A-12 ever occurred on June 21, 1968, when an A-12 was flown from Area 51 to Palmdale for storage. In all, a total of only 27 operational reconnaissance flights were made, out of a total of 2670 flights.

Titanium Goose

One of the A-12s was substantially different from the others. The fourth airframe, Article 124/serial number 60-6927, was the sole two-seat trainer. It has been variously designated as the A-12T and the TA-12, but it has become better known as the 'Titanium Goose'. A second, raised cockpit for the instructor-pilot was added to the airframe, occupying the former Q-bay. Delivered in November, 1962, its first flight occurred in May 1963. It was equipped with J75 engines and was never upgraded to J58s. The less powerful engines, coupled with increased drag from the raised second cockpit, meant that the Titanium Goose was never able to reach Mach 3. It was in fact limited to Mach 1.6 and 40,000ft. This was not a problem, however, as takeoff and landing, together with basic control of the aircraft, were more important for the trainee to learn about than high Mach flight.

The Titanium Goose was built with the original wedges in the chines, covered in triangular sheets of titanium. This was non-stealthy, but once again that was not an issue for this trainer aircraft. The aircraft survives and has been on display outside the California Science Center in Los Angeles since 2003. It is the only A-12 currently in a bare-metal finish. While this has allowed the skin to be stained and oxidized by the weather, it has resulted in a visually rather remarkable display. The Titanium Goose is also notable for being the only A-12 in which Kelly Johnson himself took a flight.

B-12

From the beginning, the Air Force had expressed an interest in the strike abilities of the A-12. So by July of 1961, a full scale mockup of the forward fuselage of an 'RB-12', also referred to as an 'R-12', had been constructed. The B-12 as then designed used an A-12 fuselage, seemingly unaltered in appearance, but with a bomb bay located in the central fuselage aft of the nose gear. The bay held a rotary launcher with four all-new advanced nuclear bombs; alternatively it could hold a single Polaris warhead.

Few details are available about the B-12 apart from the details of the bomb bay; the existence of an 'RB-12' designation implies that at least some reconnaissance capability would remain. Whether this means that it would carry recon equipment at the same time as nuclear bombs, or could swap out mission hardware between flights, is unknown. General Le May visited the mockup and was quite interested in the concept. However, political realities killed the idea: the USAF was already pouring large sums of cash down a Mach 3 bomber shaped hole called the 'North American XB-70', and the B-12 posed a threat. The idea, for the moment, went no further.

A-12CB

The A-12CB was a minor study for a major change in operations for the A-12: flight to and from aircraft carriers. The A-12 certainly seems an unlikely aircraft to operate at sea... big, finicky, fragile and with blistering takeoff and landing speeds. Numerous changes would have been required.

Known so far from only two similar, but not identical, diagrams, the A-12CB (Carrier Based) would have had sizable solid rocket motors attached to the underside. The motors were Rocketdyne RS B-202s, used for the F-104G zero length launch system.

With a total thrust of 130,000lb and a burn time of 7.9 seconds, these jettisonable take-off units would, with the aircraft carriers catapult system, have done a good job of lofting the A-12 forwards into the air. The A-12 would have required substantial internal modification both to the underside of the wings and the forward fuselage to tie these new sources of thrust into the aircraft structure. For landing, an arrester hook was fitted, which also would have required tie-in to the main structure. Presumably the landing gear would also need reinforcement.

The available diagrams show that the A-12CB was to be catapult-launched using the now-obsolete system of a bridle cable attached to two hooks located on the underside of the fuselage, straddling the centreline just aft of the inlets. The bridle cable attached to a very large catapult shuttle that seemed to serve as a support for the forward fuselage, perhaps intended to hold the nose of the aircraft down during the catapult

Lockheed A-12 'Titanium Goose'

SCALE 1/120

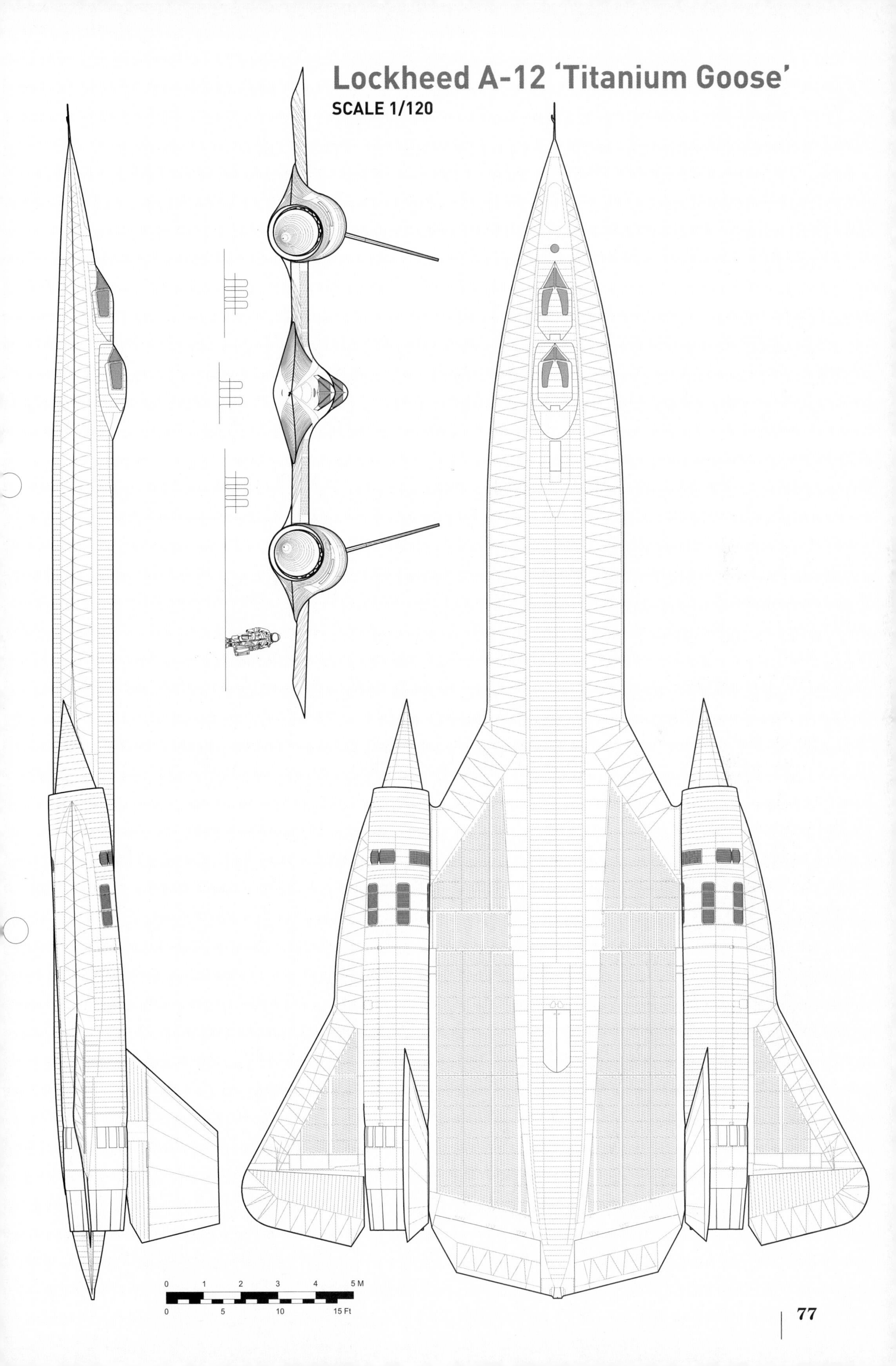

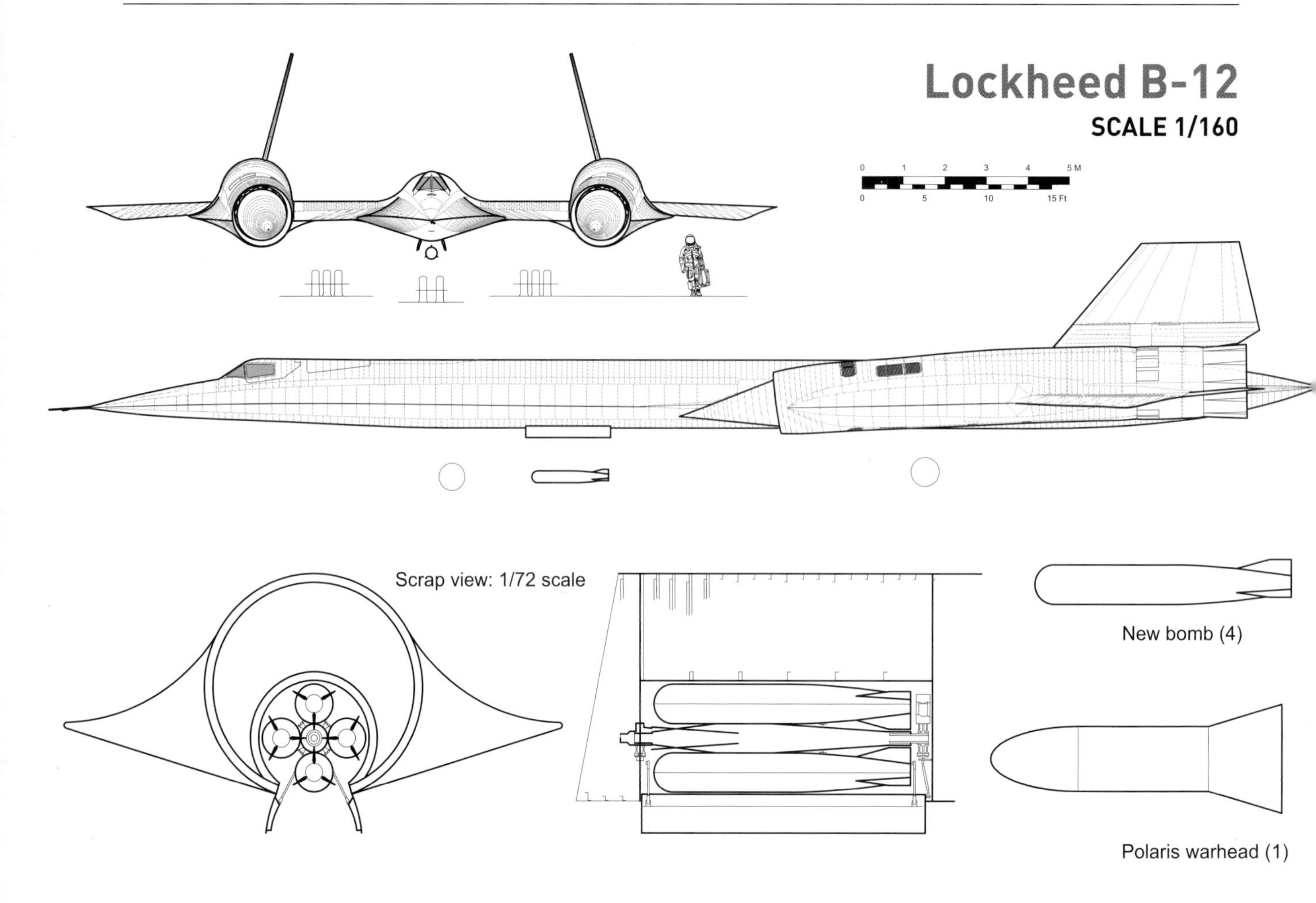

launch. With the large booster rockets firing, it's likely that substantial forces would have been in play that could have thrown the aircraft all over the deck until aerodynamic flow held it on course.

Nothing is known about what aircraft carrier or carriers the A-12CB was meant to operate from. Here it has been provisionally shown on the deck of CVAN-65 USS *Enterprise* (see p80). It can be seen that it would have been a very large aircraft for the deck of that ship, though it appears that with some effort it could have been made to fit on the ship's elevators. Whether it could have been shoved through the hangar doorway is unclear; it may have had to be partly disassembled to fit. Additionally, the deck of the carrier would likely need reinforcement, or at the very least protective coatings, to shield it from the booster rocket exhaust.

Operations would have been complex and difficult and their purpose is unclear. Presumably the A-12 would have remained a reconnaissance platform, but it is possible that the A-12 was to be a strike variant. One diagram of the A-12CB shows a crudely sketched *something* penciled in above the fuselage, but it's impossible to determine what was intended.

In any event, nothing came of the design. Dimensions, weights and performance are not available, but would presumably have been much the same as for the standard A-12.

AP-12

One of the more ambitious proposals for the A-12 was the 'AP-12' of September 1962. While the A-12 was certainly a capable reconnaissance platform, it had limitations in reach and speed. Spy satellites such as the Corona system, which began flying in 1959, could overfly virtually the entire world but they followed a predictable orbit that carried them over potential targets only infrequently. The AP-12 system would solve these problems by launching a satellite from a heavily modified A-12.

The AP-12 would carry (either underslung along the centreline or on the top of the rear fuselage) a solid-rocket booster derived from the Polaris A-3 submarine-launched ballistic missile. Takeoff would be from an airfield in Hawaii, followed by a cruise 250 to 600 nautical miles east. Launched on a south-easterly trajectory at an altitude of 80,000ft and a speed of Mach 3, the booster would send the satellite onto a single low-altitude (80 nautical miles) orbit directly over the target area, following which it would enter and be air-snatched over Johnston Island around 87 minutes after launch. Secondary recovery sites in the southwestern United States would be possible. Inertial guidance would ensure that the satellite would pass no more than five miles to the side of the primary target.

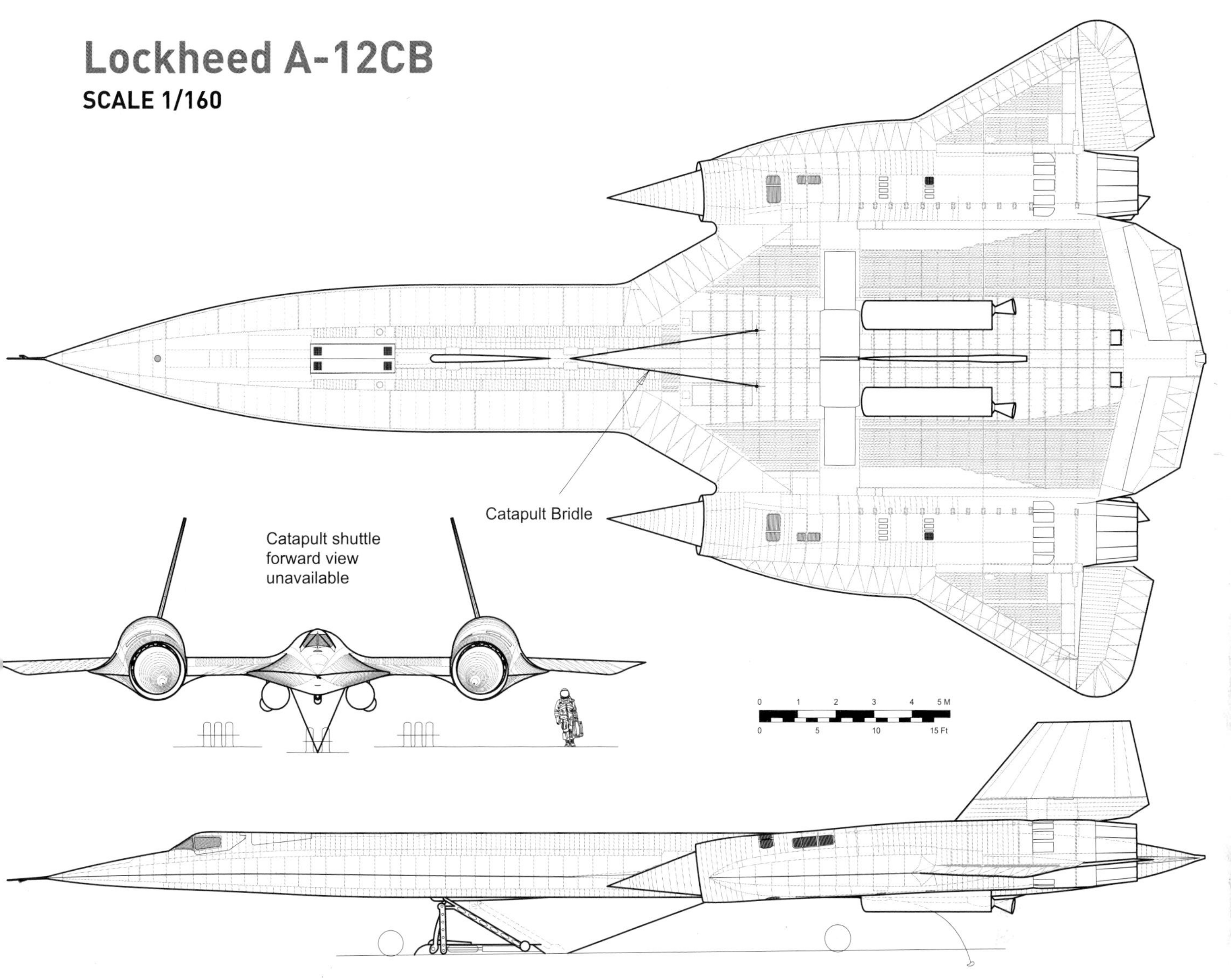

The baseline AP-12 design called for suspending the rocket underneath the aircraft. This would make separation easier than carrying it above; the rocket would more or less simply fall away. But there was a problem with geometry: the AP-12's landing gear folded up into the central fuselage, right where the rocket needed to be. So a clever yet complex and doubtless troublesome solution was envisaged. During taxiing and takeoff, the rocket and its payload would be located ahead of the landing gear; the aft aerodynamic flare and its fairing located behind the landing gear. The gear would fold up through the gap between the two components. But after takeoff the rocket would slide aft to mate with the fairing.

This would also position the booster in the best place for the weight and balance of the aircraft. After reaching the target altitude, speed and location the rocket would drop and the aft flare would deploy, stabilizing the vehicle for a few seconds as it separated from the AP-12. Upon first stage ignition the flare would be jettisoned. A new high-performance third stage would be included. The stage would be installed 'backwards' necessitating a 180° rotation of the payload and the upper stage prior to ignition. In the event of a mission abort after takeoff, the booster and satellite would be jettisoned as landing was not feasible. In the event that high-speed cruising would impart excessive heating loads onto the underslung rocket, an expendable cooling shroud could be employed.

The spy satellite itself was to be a fairly simple system compared to other satellites, given that it would have an operational life of less than an hour and a half. It consisted of little more than a telescopic Corona Mural camera, a spool of 700ft of 70mm film and a recovery system. This would provide for coverage of 180,000 square miles of surface, a strip 92 nautical miles wide with a medium-contrast resolution of 5ft. Exposed film was fed directly onto a reel within a cylinder/flare re-entry vehicle. At the end of the flight the entry vehicle would separate from the main spacecraft and a small solid rocket motor would impart a 1300ft per second deceleration, leading to prompt entry. The capsule would deploy a parachute and, optimally, would be recovered in the air though it was designed to survive splashdown.

CVAN-65 USS Enterprise

SCALE 1/1800

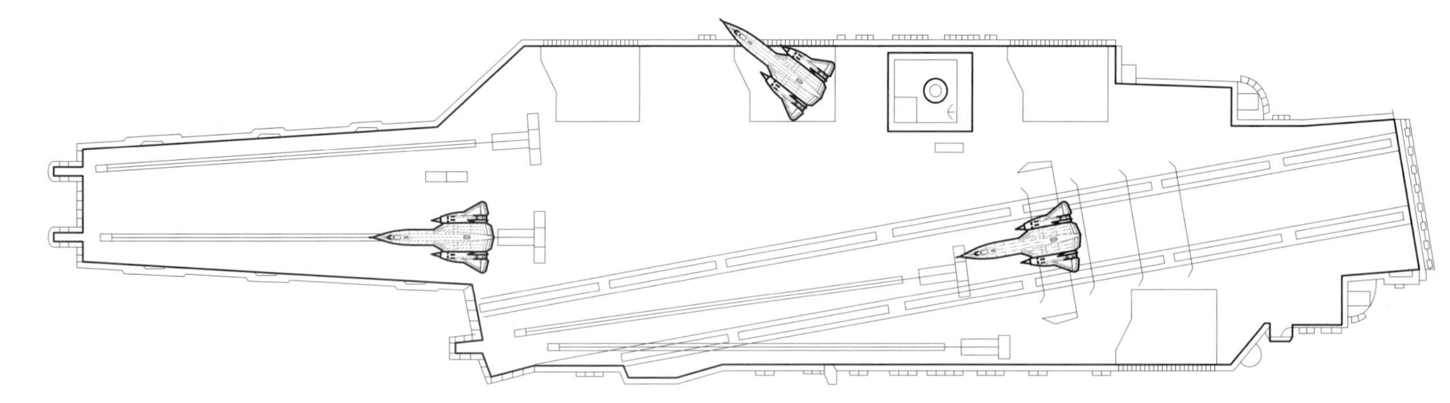

An alternate configuration was studied, though not preferred. Rather than carrying the missile underneath the AP-12, it was carried on top. There was no need to split the vehicle to provide landing gear clearance, but it required the rocket to slide all the way aft on rails (pulled by a parachute), shifting the CG of the aircraft dangerously while it did so.

Provision was to be made for a second crew member but diagrams do not give indications of windows or the slight bump that characterized the M-21 carrier.

M-21/D-21

Two of the A-12 airframes were built from the beginning for a very different role. As the programme was developed, the CIA became concerned about the risks of sending a manned vehicle – even one as advanced as the A-12 – over enemy territory. The loss of a pilot and advanced technology over 'denied' territory could prove not only technologically devastating but also politically embarrassing, as the shootdown of Francis Gary Powers had amply demonstrated. So they requested that Kelly Johnson and his team study unmanned drones to be launched from the back of the A-12 as an alternative.

By 1962 Johnson had done at least a few studies on using the A-12 as a carrier for a QF-104. There is little more to go on than the simple description of the idea, so it's uncertain how changed the QF-104 would have been compared to a standard F-104. The diagram included here should be considered wholly provisional. However, at first glance the concept seems feasible enough. But it would be strange to use a more or less stock F-104, capable of reaching about Mach 2 with a ceiling of 50,000ft, launched from an aircraft capable of flying more than 50% higher and faster. It's entirely possible that the J-79 turbojet would have been replaced by a ramjet, with aerodynamics and structure changed to permit higher speeds and altitudes.

The A-12/QF-104 idea apparently did not grab the CIA's interest and an all-new, highly optimized drone was wanted instead. Lockheed started off with a design that is currently not clearly known but was apparently essentially a subscale A-12, having a delta planform with three vertical stabilizers. Being of roughly the same configuration as the A-12 but smaller, the radar cross section would be commensurately smaller, leading to a vehicle more difficult to detect. Initially the drone was designated the AQ-12, 'Q' being a standard type designator for unmanned aerial vehicles. The programme was codenamed 'TAGBOARD' by the CIA in September 1962. Lockheed's feasibility study was completed in January 1963 and the following February the company was given the go-ahead to produce 20 of the new drones. The project was to be carried out with the highest of security classification.

Early on, the idea was raised to use the drone as a strike vehicle. However, the payload capability of the AQ-12 would be minimal, only around 250lb, shoved into the volume of a camera. Nuclear warheads of the time would have struggled to fit in such a package, and a conventional warhead of that weight would have been spectacularly wasteful of the delivery system.

The design soon morphed into a blended single-ramjet design that resembled the A-12, but was clearly not merely a reduced-scale A-12. This design had a metallic central structure with the chines and large portions of the wings and vertical tail to be made from composites. The wings were of a modified delta planform with sizable rounded chines that served a canard-like aerodynamic function. The wings had substantial anhedral. Control surfaces were few: a rudder on the vertical stabilizer and elevons on the wing trailing edges.

This early configuration appears in the form of photos of a full scale mockup and it was largely like the vehicle that would actually be built. However, there

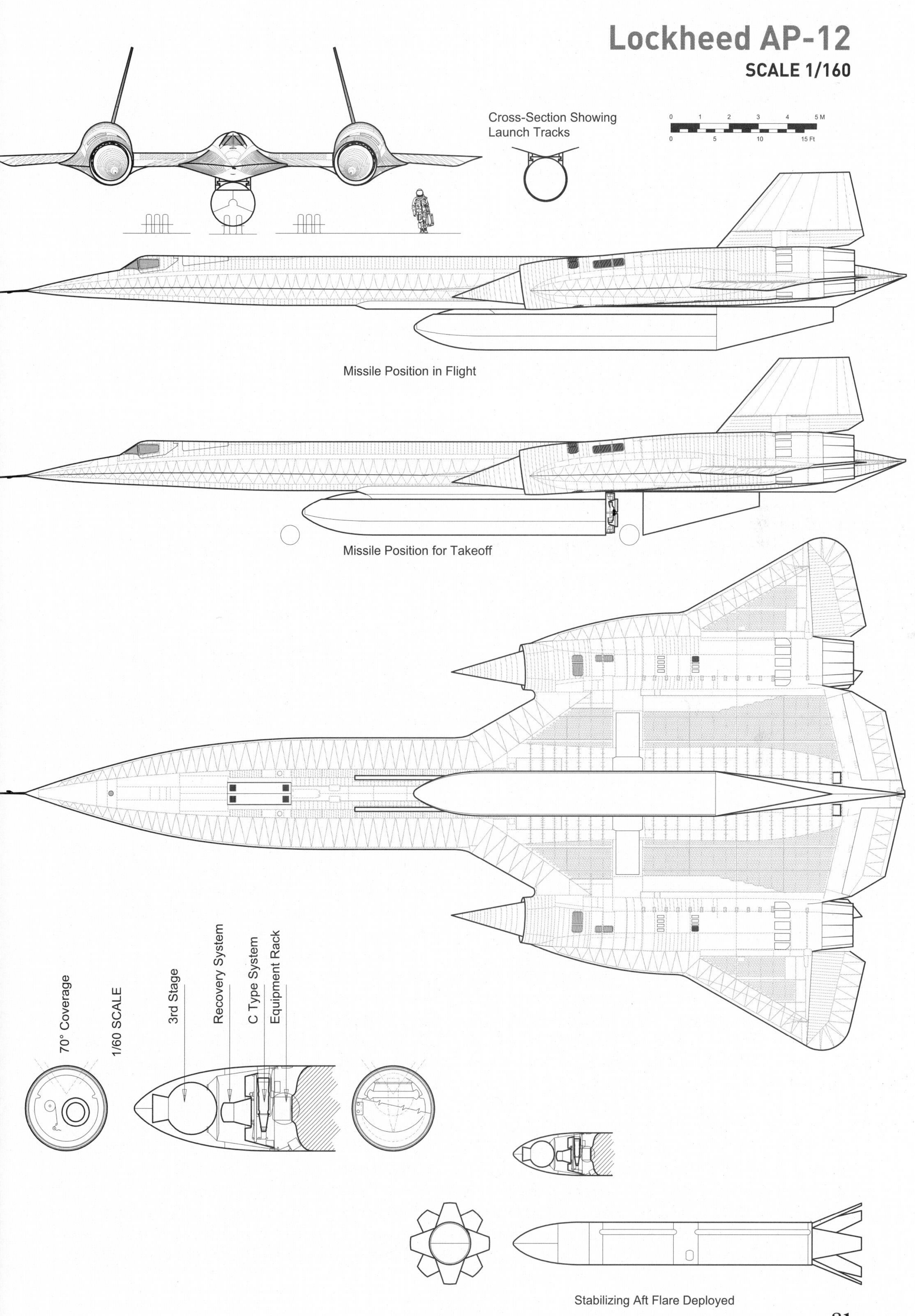
Lockheed AP-12
SCALE 1/160
Cross-Section Showing Launch Tracks
0 1 2 3 4 5 M
0 5 10 15 Ft
Missile Position in Flight
Missile Position for Takeoff
70° Coverage
1/60 SCALE
3rd Stage
Recovery System
C Type System
Equipment Rack
Stabilizing Aft Flare Deployed

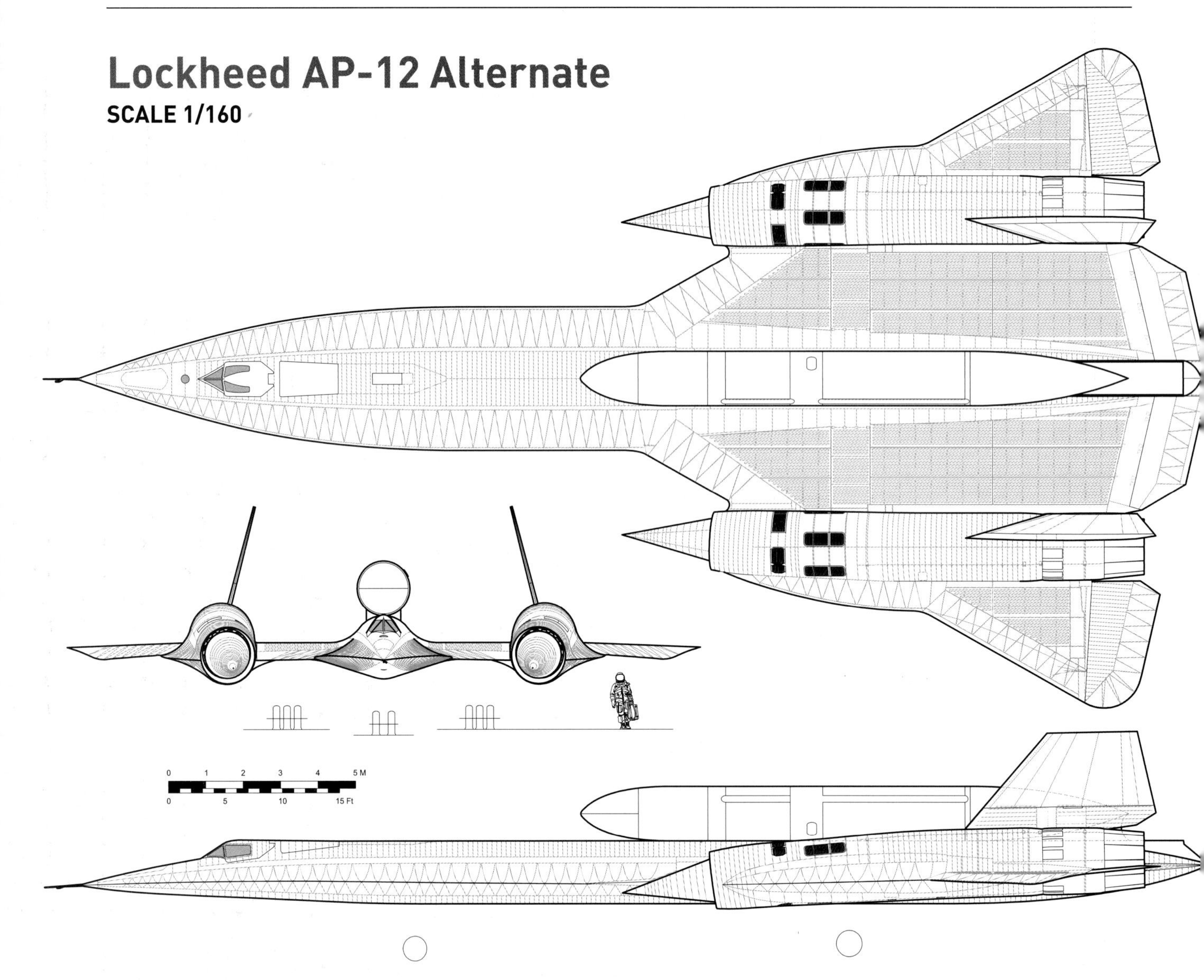

were a few differences in contours and details. A disk-like fairing was held above the fuselage, presumably containing a command-receiving antenna.

The fuselage was a semi-monocoque structure of rib and skin construction, the bulk of which was titanium with a few parts made of stainless steel. The wing structure was again of titanium sheet covering ribs that were integral with the fuselage structure. The chines, wing leading edges and elevons were made from a honeycomb covered in a silicone-asbestos laminate, very likely the same makeup as the chines of the later A-12s. 'Plastic' elevon fences – small vertical fins – were located just inboard of the elevons on the upper surfaces of the wings. The control surfaces were hydraulically actuated.

To prevent confusion, in October 1963 the A-12 'mothership' was designated 'M-21', M for 'Mother' and 21 being simply a reversal of the 12 to prevent confusion. The AQ-12 was redesignated D-21, D for 'Daughter'.

The D-21 was powered by a single Marquardt RJ-73 Model MA20S-4 ramjet. This was derived but greatly modified from the Marquardt RJ43-MA-3 ramjet engine used on the IM-99 Bomarc surface-to-air missile. The engine was given a larger exhaust nozzle with a greater expansion ratio for improved high-altitude performance. A number of other design changes were made to account for the higher cruise altitude and much longer flight duration. The engine itself was in the extreme rear of the fuselage, fed by a long and somewhat sinuous duct from a circular inlet – equipped with an inlet spike, fixed, unlike the A-12's – in the nose. The D-21, like the A-12, used triethylborane to ignite the ramjet, both during the initial engine startup and in the event of a flameout. The fuel, designated PSJ-100B, was a low aromatics, low volatility mixture that sounds much like JP-7.

The duct bulged upwards near the front in part to provide room beneath for the surveillance system – the 275lb Aerial Reconnaissance Camera System Model HR-335 with a 24in focal length f/5.6 camera. This created 9x9in images on one 4500ft-long roll of film, enough for 5600 frames, with a resolution of 2ft at an altitude of 85,000ft. It could operate in one of two modes – Mode 3 and Mode 5. In Mode 3, assuming a

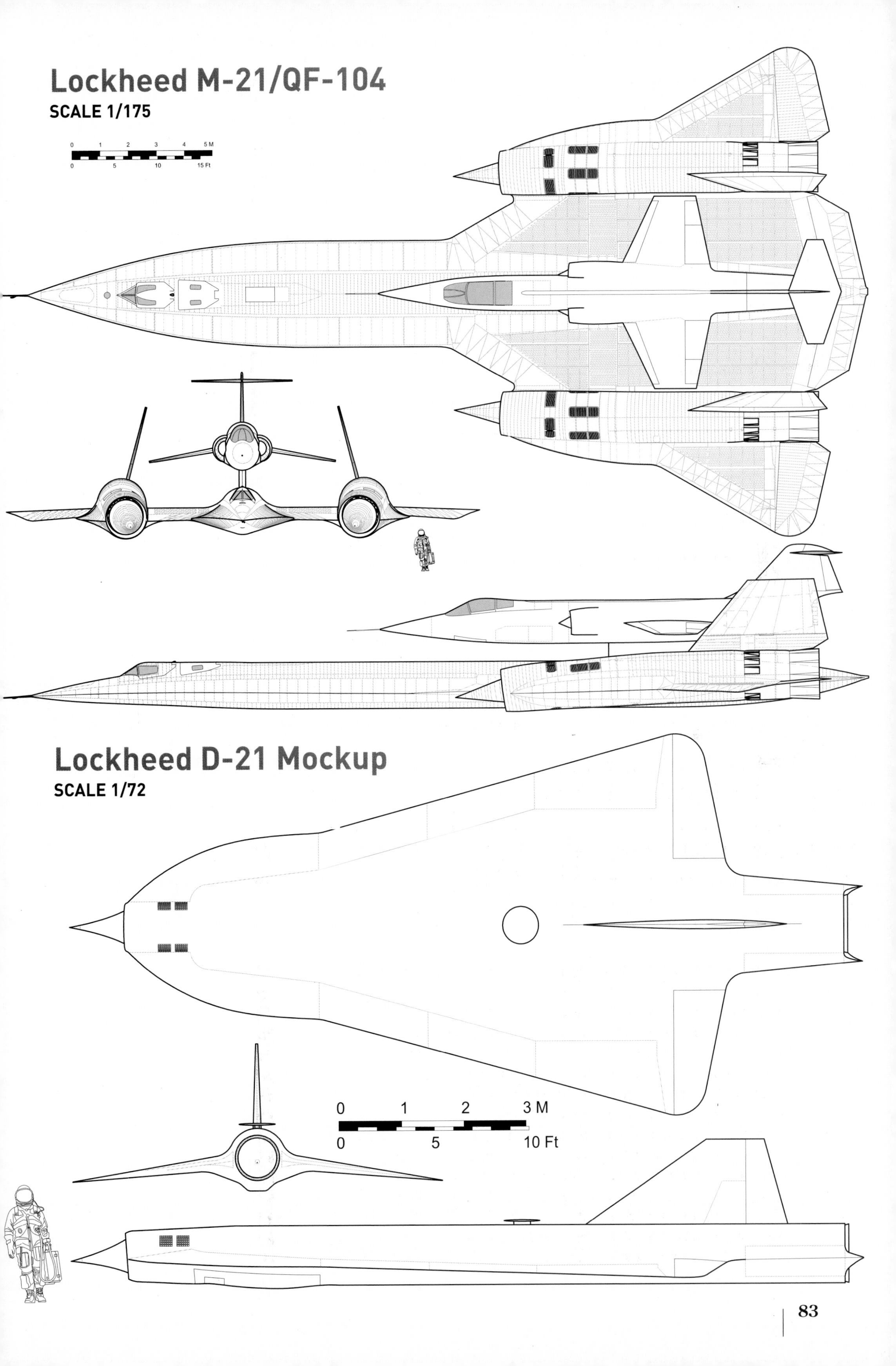

Lockheed M-21/QF-104
SCALE 1/175
0 1 2 3 4 5 M
0 5 10 15 Ft
Lockheed D-21 Mockup
SCALE 1/72
0 1 2 3 M
0 5 10 Ft

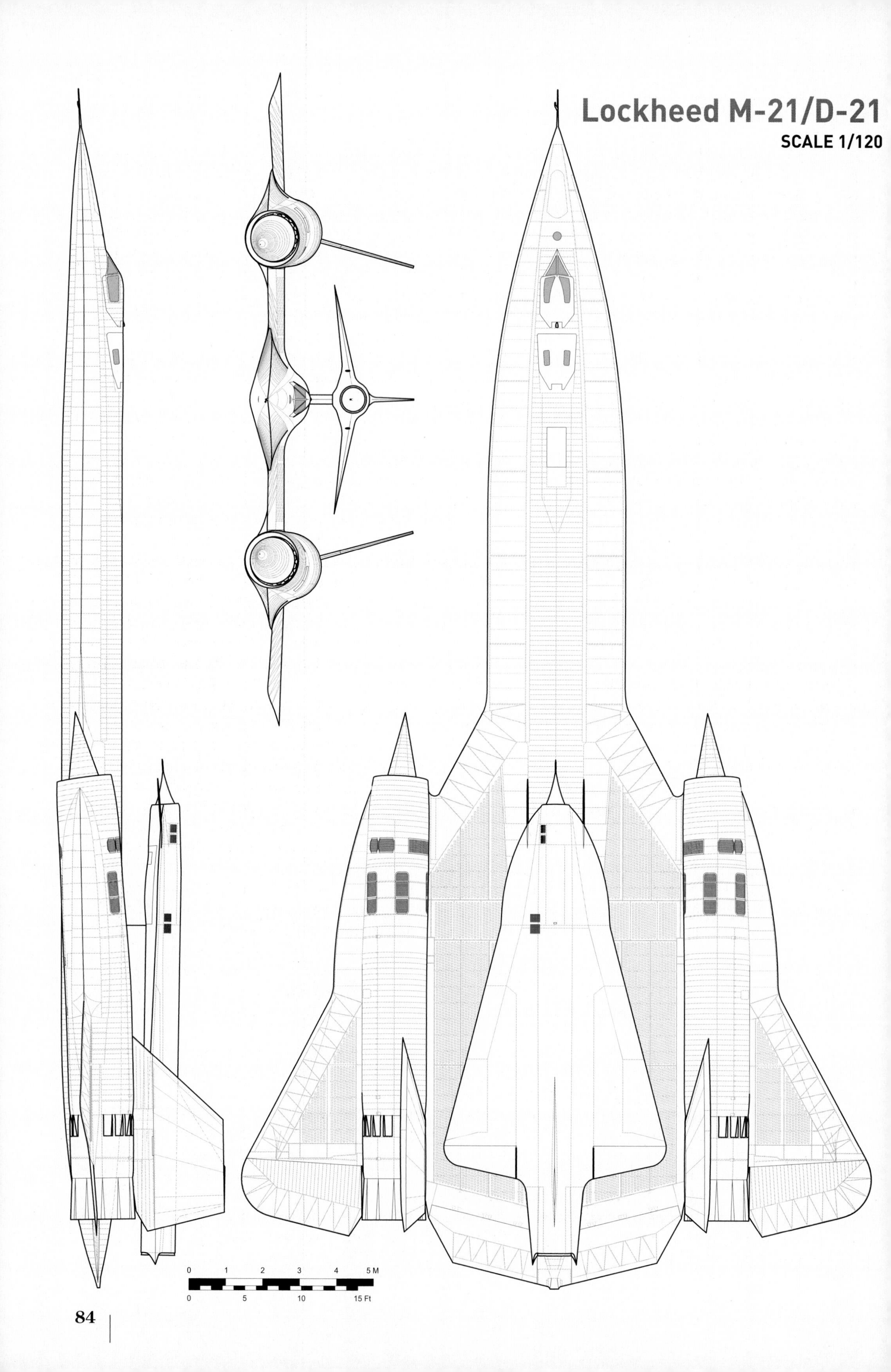
Lockheed M-21/D-21
SCALE 1/120
0 1 2 3 4 5 M
0 5 10 15 Ft

flight altitude of 80,000 to 95,000ft the width of the images would be from 15 to 18 nautical miles for a coverage of 3700 nautical miles. Mode 3 would take images at three pointing angles... 19° left, vertical and 19° right. In Mode 5, the width of the images would be 26 to 31 nautical miles for a coverage of 2780 nautical miles. Overlap between sequential frames would be 60% for both modes. Mode 5 would take images at three pointing angles... 36° left, 19° left, vertical, 19° right and 36° right.

The D-21 included a Honeywell MH-390(D) inertial navigation system capable of guiding the D-21 to up to 16 destinations. Position could be updated while on the M-21 up to the moment of launch, with a maximum initial position error of +/- 1.7 nautical miles using the M-21's stellar tracker. While the D-21 was self-navigating, it was capable of receiving signals for the purpose of fuel shutoff, self-destruct and jettison of the camera system for recovery.

The D-21 was an expendable aircraft. It had no recovery system – no landing gear and no parachute. At the end of each mission it was to plunge into the sea for complete destruction. However, the vehicle had a hatch on the lower side designed to be jettisoned. Attached to this hatch was the camera system including all of the exposed film, along with the inertial navigation system and the automatic flight control system. Additionally, the hatch had a 'Sarah' radio beacon, an X-band transponder and an automatically deployed parachute.

In an ideal situation the hatch would be jettisoned at an altitude of 60,000ft over a pre-determined spot of the sea. There a specially modified C-130 would be waiting to air-snatch the parachute and its payload; the hatch would be brought on board and promptly flown to a base for film processing. In the event the C-130 failed to capture the hatch and parachute while in flight, it was designed for floatation, with all of the equipment in a water-tight package. The hatch and associated structures were largely made of titanium.

The M-21 received substantial modification in order to carry the D-21. The Q-bay was replaced with a compartment for a second crew member, the launch system officer. The crew compartment did not quite fit within the confines of the Q-bay, resulting in a subtle outward bulge of the upper fuselage. The M-21 was also fitted with a long, low pylon on the upper rear fuselage for carrying the D-21; the pylon included systems to jettison the D-21 as well as fuel transfer systems. This latter feature would come in handy as, during the D-21's troubled development, it was found that the additional thrust of the ramjet engine of the D-21 was needed during the run-up to Mach 3. The braking parachute compartment on the upper side of the fuselage of the M-21 was located further aft than on the A-12 due to interference from the pylon.

Wind tunnel testing showed that separating the D-21 from the M-21 at Mach 3 would be a challenge. The D-21 would have difficulty penetrating the shockwave generated at the nose of the M-21; it was found that in order to successfully separate the M-21 would have to perform a diving 'pushover' manoeuvre. With only inches to spare between the D-21s wingtips and the inward-canted rudders of the M-21, precision and reliability were key.

Flight testing of captive D-21s began in December of 1964 and proved occasionally difficult. In one flight the elevons of the D-21 experienced excessive flutter and were damaged. In order to minimize drag, the inlet of the D-21 was covered in a frangible cone and the exhaust nozzle given an aluminum fairing. When separated, the nose cone sent shards into the engine and damaged the leading edges of the D-21's chines. Both the nose and tail fairings were abandoned. While this increased drag, it did free up the ramjet to run while still attached to the M-21.

The first launch of a D-21 occurred in March of 1966. The drone successfully separated and flew 120 miles. Two more successful D-21 flights followed, in April and June, flying 1200 and 1600 nautical miles respectively. But in July 1966 the fourth launch failed catastrophically. The problems Lockheed found during wind tunnel testing of the D-21 having difficulty penetrating the M-21's shockwave resurfaced. Shortly after separating the D-21 simply bounced off the shockwave, came back down and stuck the M-21, causing the aircraft to break apart in flight. Both crew ejected safely and parachuted to splashdowns in the ocean, but the launch systems officer drowned. With the loss of one of only two M-21s, the decision was made to abandon the idea of launching the D-21 from a Mach 3 manned aircraft.

This was not the end of the D-21, however. In early 1966, Kelly Johnson had suggested using the B-52 as a carrier aircraft for the D-21. In order for that to work, the D-21 would be equipped with a large solid rocket booster to get it up to speed and altitude. The idea was approved under the code name 'Senior Bowl', and two B-52H bombers were provided for modification.

Lockheed built four pylons to carry the drone, redesignated D-21B (the D-21s already built were modified to have new attachment points in their upper surface). The pylons were similar in appearance to the pylon used to carry the X-15, and would attach to the inboard hardpoint already built into the B-52H for the carrying of Hound Dog and other payloads. The B-52H could, and occasionally did, carry two D-21Bs and their boosters, but only launched from the left pylon; the right pylon was used solely for transportation of the D-21B.

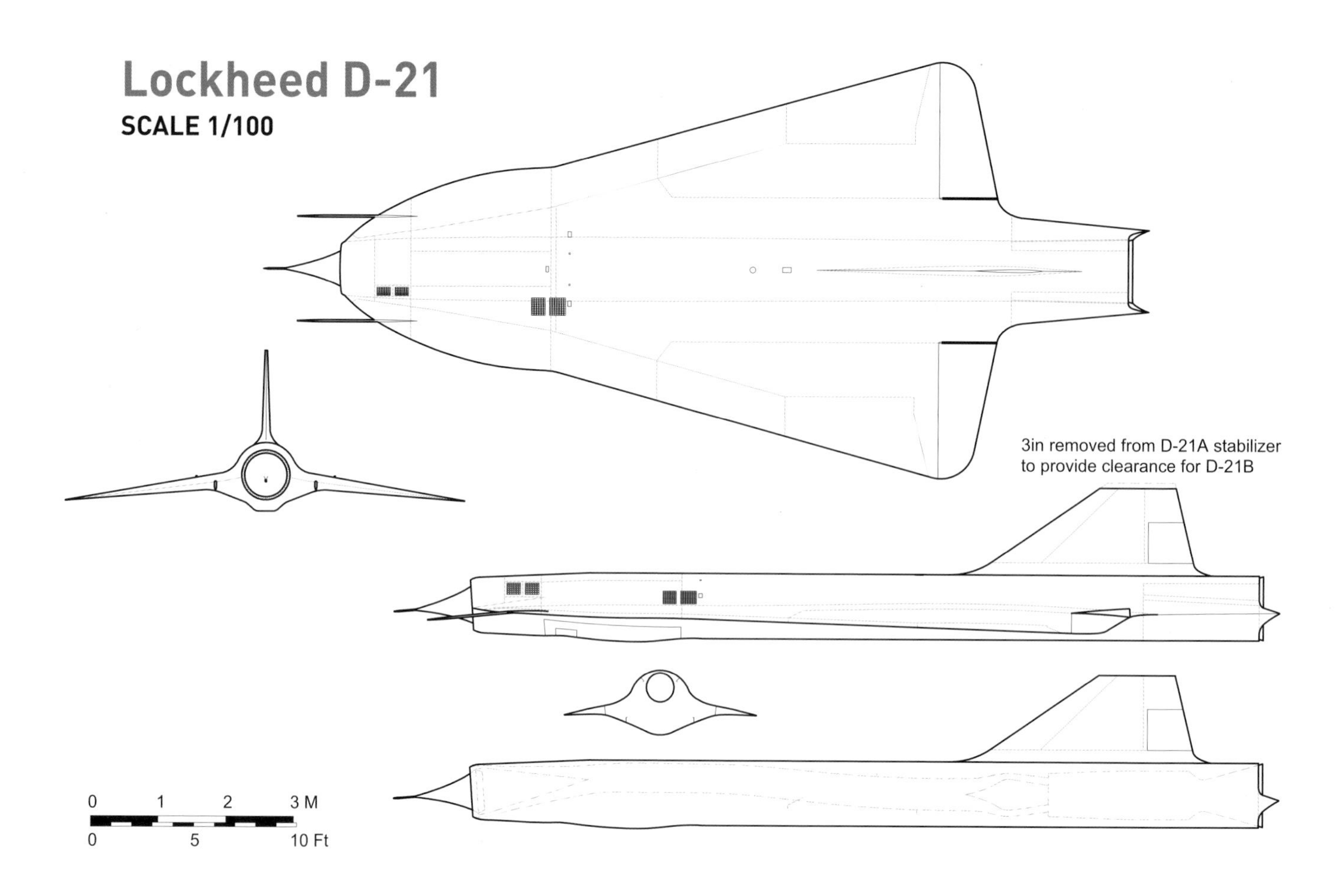

The B-52H flight deck was modified to include two launch control officer stations, taking the stations formerly occupied by the electronic warfare officer and the gunner. A stellar inertial navigation system was added along with telemetry and redundant command communications systems. A periscope above the flight deck could be used to observe both D-21B stations in flight. The launch control officer received telemetry data regarding the position and speed of the D-21B for the first 10 minutes of flight, during which time he could order the D-21B to self-destruct if necessary. After that the drone would be out of range of the B-52; it would automatically self-destruct if its altitude fell too low.

The booster for the D-21B consisted of a Lockheed Propulsion Company A-92 'Avanti' solid rocket motor, a fairing assembly (including a nose, pylon to connect to the D-21B and a tail unit with a deployable ventral fin) and a ram air turbine in the nose to generate power. The overall length of the rocket motor itself was 422in with a main tube diameter of 30.16in and a diameter over the flanges of 32.62in. With an overall weight of 12,500lb, it could produce an average thrust of 27,300lb for about 87 seconds. The weight of the booster including the rocket motor and external structures (but not ballast) was 13,000lb. The propellant was low burn rate Polybucarbutene R, a high energy propellant with 68% ammonium perchlorate and 17% aluminium powder.

The booster would burn out at an altitude of about 80,000ft and above Mach 3.3. After that, the drone would continue on its mission, initially climbing to a little short of 90,000ft before settling down to around 82,000ft. Through the course of the mission, as fuel was burned off, the drone would maintain a speed of Mach 3.3 but climb to around 94,000ft. At the end of the mission it would quickly descend, jettison the camera hatch and then self-destruct. The circular error probability of the D-21B at the end of its mission was 4.7 nautical miles.

The first launch of a D-21B from a B-52H was carried out in November 1967. The booster took the drone to altitude, but the drone failed, coming down about 150 miles away. A number of aborted launches and unsuccessful test flights followed; it was not until June 1968 that a truly successful flight occurred. An altitude of 90,000ft, a speed of Mach 3.3 and a range of 3000 miles were achieved, along with camera recovery. The ramjet flamed out during programmed course turns but the TEB system re-ignited the engine and the aircraft continued. More test flights, successful and unsuccessful, followed. Finally an operational mission over China was launched in November of 1969; on this flight the D-21B simply

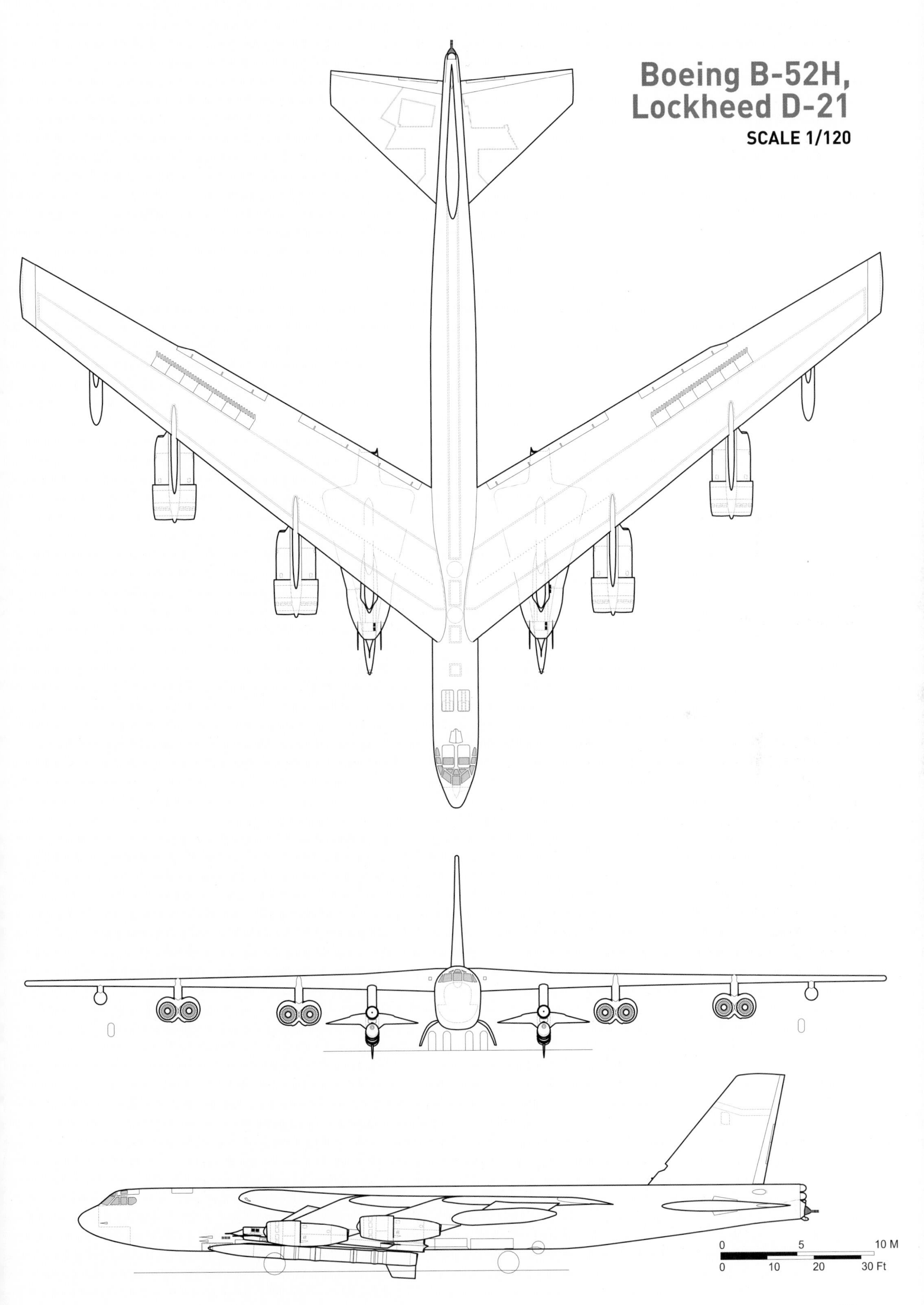
Boeing B-52H,
Lockheed D-21
SCALE 1/120
0
5
10 M
0
10
20
30 Ft

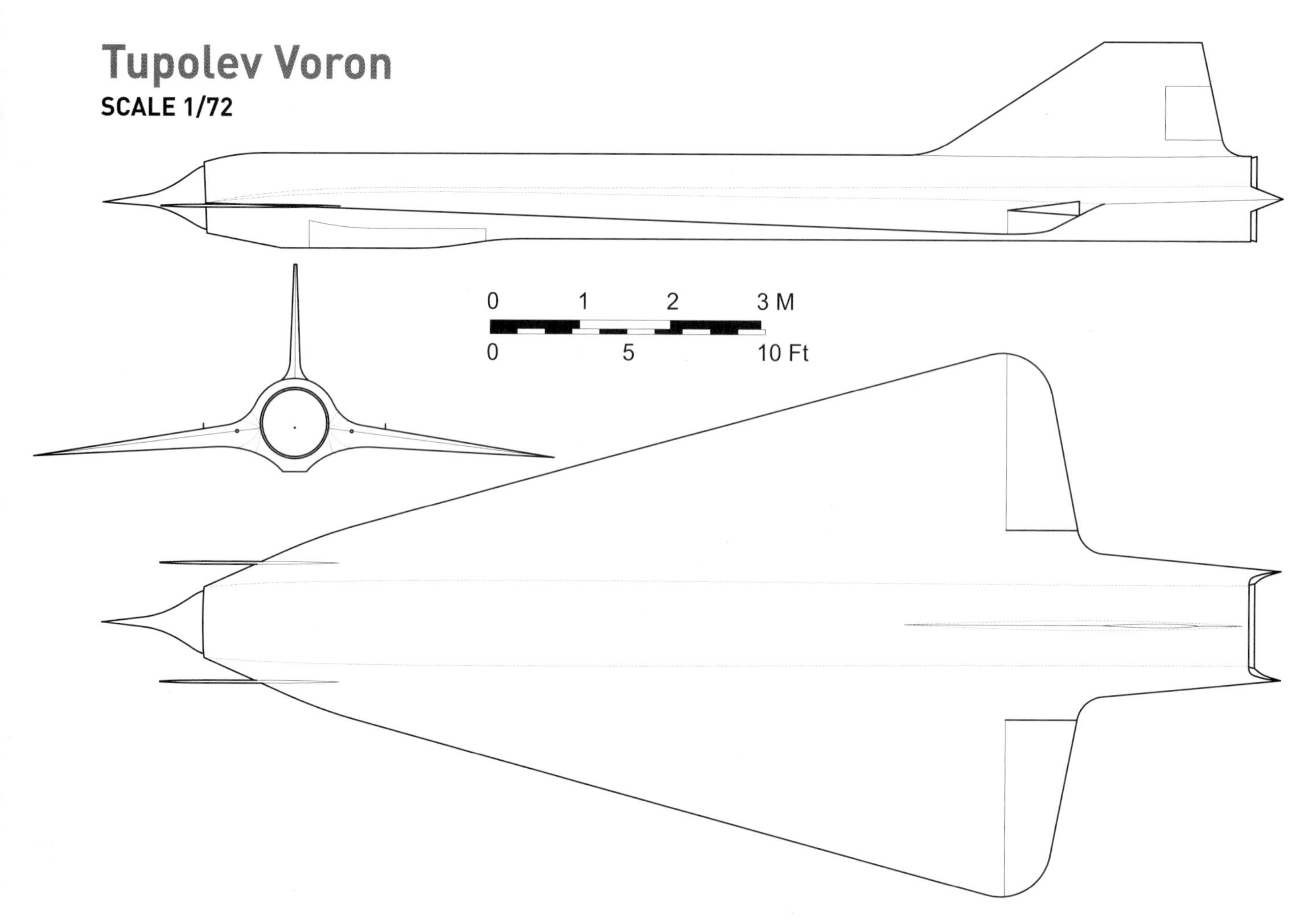

vanished, likely due to navigation system problems. The second operational mission was more than a year later in December 1970. The drone flew the programmed route successfully, but the hatch was not recovered. The third operational mission was in March of 1971. Again the flight itself was successful, but the hatch was damaged during recovery and the photos were ruined. The fourth and final operational flight was also in March of 1970; the drone was lost over heavily defended enemy territory.

The D-21 programme consumed many years, many manhours and many dollars without a single useful reconnaissance photo to show for it. And while the D-21 was struggling on, spy satellites were becoming ever more capable and the usefulness of the programme fell into serious doubt. In July of 1970, the D-21B programme was cancelled, the surviving airframes ordered into storage.

The end of the TAGBOARD and Senior Bowl programmes were not the end of interest in the D-21. One of the more remarkable examples of a derivative project was the Tupolev 'Voron' ('Raven'). The first D-21B flown over China, so far as the operators of the vehicle knew, simply vanished. But it didn't. It would seem that the navigation system failed to turn the craft around after photographing the Chinese nuclear site at Lop Nor and it continued on north into Siberia before running out of fuel and crashing. It was recovered by the Soviets and studied. And, in grand Soviet tradition, in March 1971 the decision was made to copy it. This task was given to the Tuplolev Design Bureau. The resulting design could be easily mistaken for a standard D-21, so long as one paid no attention to the Cyrillic.

The Voron was visually almost identical to the D-21, but there were a few differences. Most apparent was a change to the chines... on the Voron they did not curve outwards beyond the main leading edge line of the wing. The inlet spike was contoured differently, with a more pronounced concave curve; and the camera package fairing was a little different. Internal differences were more pronounced. The Marquardt ramjet engine was not slavishly copied, but was replaced with a new design, seemingly a modification of the RD-012 from the 'Burya' cruise missile.

The Voron was not meant to be launched like the D-21 from atop an M-21-analogue. This is not surprising as the M-21/D-21 should have been unknown to the Soviets, while the D-21B wreckage on-hand showed evidence both of an underslung booster and overhead carriage. The Soviets therefore designed a new booster for the Voron (shorter and fatter than the Lockheed design) and carried the drone beneath, initially, a Tupolev Tu-95KD missile carrier. It was apparently expected that the Voron would eventually be carried by the then under-development Tu-160. At the time the Tu-160 was envisioned as a derivative of the Tu-144 SST. The supersonic bomber would have launched the Voron without need of the rocket booster, but unlike the M-21, the Tu-160 would have dropped the Voron from beneath.

While lessons seemed to have been learned in the areas of materials and high-speed design, the Voron was not built. As with the D-21, it had difficulty competing against spy satellites, and the Soviets cancelled the programme before construction could begin.

After the Senior Bowl programme was shelved, the 17 remaining D-21Bs were mothballed. For years they sat unused and largely forgotten but in 1999 an opportunity for revival came in the form of a NASA programme. In the 1990s there was a glut of space launch notions, including a bubble of commercial small reusable launch vehicles (which this author got in on). Included in that was a brief fascination with certain specific technologies, one being the Rocket Based Combined Cycle (RBCC) engine. This type of engine has been studied and occasionally built since the 1960s, holding theoretical promise as the propulsion system of 'orbital aircraft'. The engine looks like a ramjet, but embedded within it are a number of rocket engines. The rocket engines work with the ramjet duct and fuel injectors in various ways at various flight regimes to produce an engine that, theoretically, can be used from runway to orbit.

From Mach 0 to Mach 3 or so, the engine works as an air augmented rocket (AAR). The rocket engines fire, producing thrust; but the rocket exhaust interacts with the surrounding air and, through shear forces, entrains and accelerates a substantial mass of that air. Any rocket exhaust will do this, but it plays no

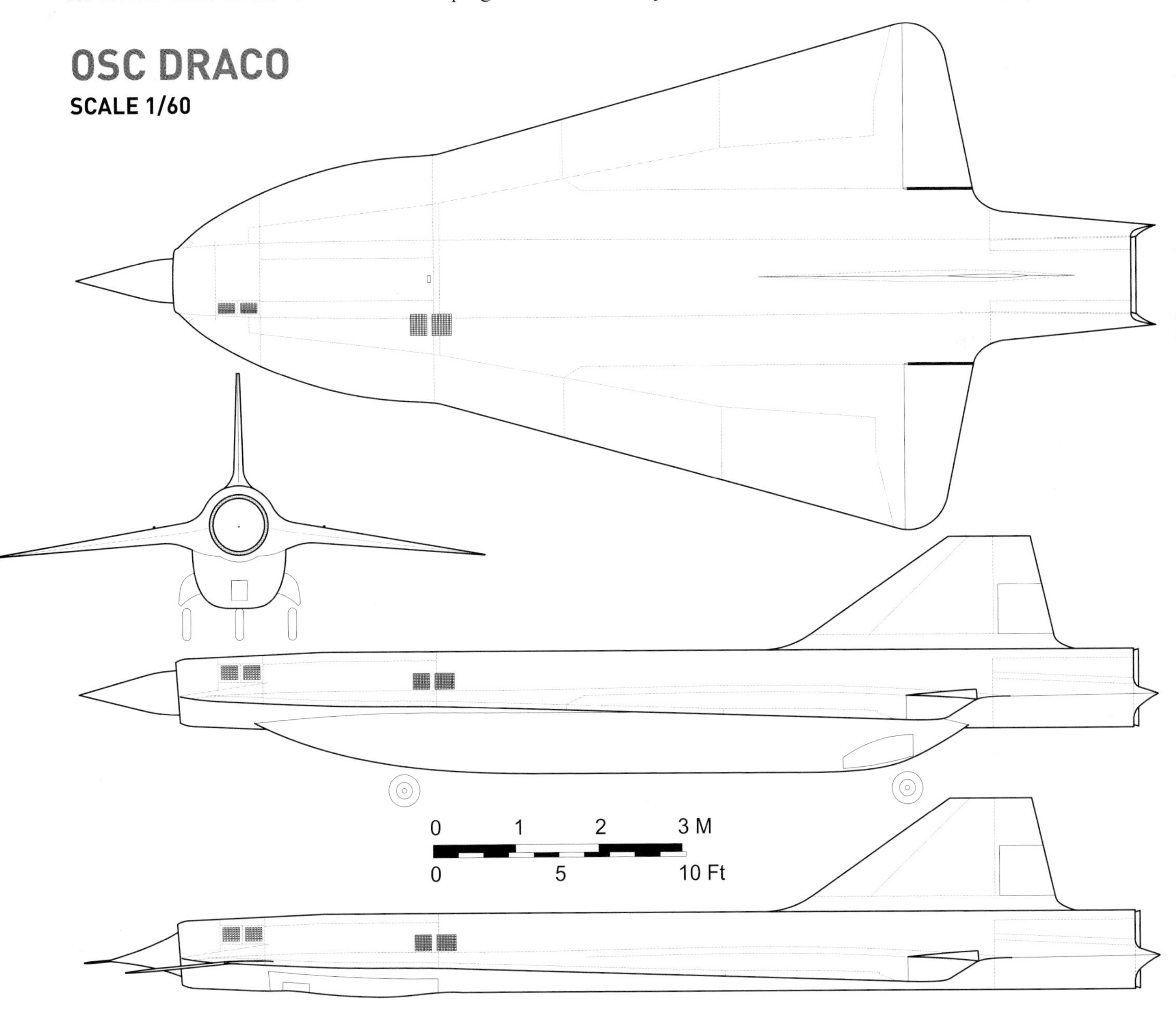

meaningful role as the effect occurs downstream of the nozzle, well behind the vehicle. An AAR, though, has the rocket engines within a ramjet duct. The air, drawn in through the inlet, is shoved aft through the nozzle. The velocity of the rocket exhaust is greatly reduced, but the total mass flow rate is greatly increased. Done right, this alone will increase thrust at low speed.

Additionally, the air mass flow and velocity can be high enough that injectors upstream of the rocket engines will add fuel to the flow, fuel that will combust and make the engine function as a ramjet or afterburner. As the engine reaches Mach 3, the rockets are shut down, allowing the engine to function purely as a ramjet. Above Mach 6, the ramjet is shut down while the rockets are turned back on; the engine then functions as a rocket engine all the way to orbit if need be. That, at least, is the theory. Reality, though, has a way of being irritatingly detail-oriented, and such engines tend to be finicky, complex, expensive and, as this author discovered while working on an air augmented ramjet programme at the time, often perfectly willing to produce much less thrust than the math would suggest.

Still, decades of work had shown the concept to have promise, so at the end of the heady 1990s – a decade that had seen the National Aero Space Plane programme, the Delta Clipper, the X-33, the VentureStar, the X-34 and the X-Prize – NASA-Marshall Space Flight Center wanted to revive the idea. More, they wanted to go further than had been gone before by actually flying an RBCC engine. The DRACO (a somewhat strained acronym for Demonstration of Rocket and Airbreathing Combined-cycle Operation) programme was to build and fly such an engine and, uncharacteristically for a NASA programme, the plan was not to design a bleed-edge airframe to go with it. Instead the idea was to use one of the mothballed D-21B airframes; three being stored at the time at the NASA Dryden Flight Research Center.

While the D-21 was intended to be an expendable vehicle, it was well enough built that recovery and reuse was possible. To do this a recovery system would be needed; the most straightforward approach would be landing gear. But the D-21 had never been wind tunnel tested at low speed, since it was never meant to fly slowly. The landing characteristics of the D-21 would need to be proven.

NASA asked the American aerospace industry to submit proposals on how to modify the D-21B for the mission. Two submissions are known reasonably well: unsurprisingly, Lockheed considered the idea, as did Orbital Sciences Corporation. Lockheed's concept would have externally changed the D-21 very little. A deployable nose wheel would have been added to what was previously the camera compartment, while a deployable skid would be added under each wing. The RBCC engine would fit within the confines of the D-21s ramjet, and the fuel (JP7) and oxidizer (liquid oxygen) would be stuffed into conformal tanks that fit within the existing internal tanks of the D-21.

OSC, on the other hand, added an underslung conformal fairing containing new propellant tanks as well as deployable main landing gear. The nose gear would again deploy from the former camera bay. Both designs replaced the fixed inlet spike of the D-21B with a new slightly longer spike capable of forward/aft translation. Both companies suggested replacing the leading edge composites with new materials capable of taking higher temperatures; radar cross section reduction was of course not a consideration.

Neither company seems to have taken the study to a point of great detail, but rather only to preliminary feasibility studies. Nothing seems to have come from the DRACO concept.

One last somewhat tangential connection exists for the D-21. In 1974, DARPA began a programme to develop a practical radar-invisible combat aircraft. This was inspired by a small Lockheed project from 1973 called 'Harvey', after the invisible (likely imaginary) rabbit from the movie of the same name. With DARPA providing direction and funding, the Lockheed Skunk Works initially pursued several approaches to achieving true stealth. In May 1975, preliminary designs were produced at the Skunk Works for small stealth technology demonstrators that were based on the D-21 shape. The D-21 was, after all, designed to be stealthy, using the best configuration and materials techniques of the early 1960s.

Two similar such designs were produced and called 'Little Harvey', sharing a common planform. But in profile and front view the designs differed somewhat, though still clearly drawing from the D-21. Both had underslung inlets, substantial anhedral and D-21-like vertical stabilizers; both had a cockpit entirely submerged within the fuselage. The windscreens were not defined, at least on the available rather simple diagrams. The windscreens shown here are therefore speculative. To all appearances, the Little Harvey designs, which are known almost solely from two drawings, received only cursory examination; in the end Lockheed went with the facetted approach of the 'Hopeless Diamond' which led to the XST and eventually the F-117.

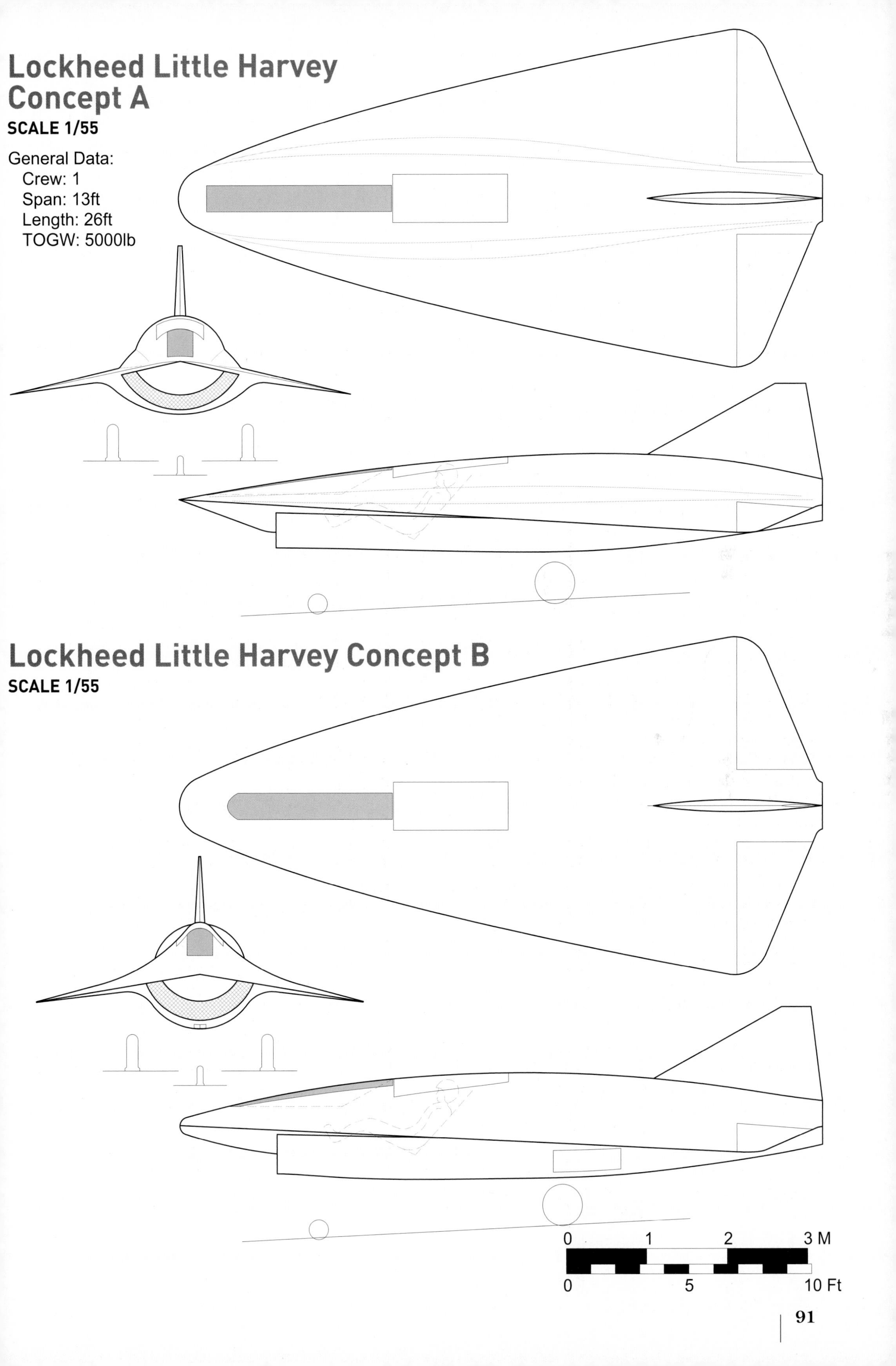
Lockheed Little Harvey Concept A
SCALE 1/55
General Data:
Crew: 1
Span: 13ft
Length: 26ft
TOGW: 5000lb
Lockheed Little Harvey Concept B
SCALE 1/55
0 1 2 3 M
0 5 10 Ft

The Mach 3 Interceptor

Those who fly, design and procure military combat aircraft have always wanted them to go faster – an impetus most succinctly expressed in the concept of the interceptor. Beginning with aircraft in the Second World War – especially the German Messerschmitt Me 163 Komet – the interceptor concept began to crystalize: launch on a moment's notice, go in a straight line as fast as possible and unload ordnance onto marauding enemy bombers. Interceptors are not generally intended to dogfight or drop bombs, theoretically simplifying their design and operation.

The dramatic post-war increases in speed made possible by rapidly improving turbojets and afterburners led to the United States Air Force focusing on interceptors over fighters, in particular with the Convair F-102 and F-106. These planes were intended primarily to defend the continental United States from Soviet long-range strategic bombers such as the Tupolev Tu-95. The F-102, which first flew in 1953, could reach a speed of Mach 1.25; the F-106, with a first flight in 1956, could reach Mach 2.3. But even those speeds were not as fast as the Air Force would have liked.

In 1949 the Air Force issued a Request For Proposals for a "1954 All Weather Interceptor". This would, as the name suggests, be an aircraft to enter service in 1954, having the latest and greatest technologies. The entire American aviation industry was interested in this proposal, including the Republic Aviation Corporation. The company's pitch, tendered to the Air Force in January, 1951 and given the Republic designation AP-57, was a sleek and seemingly simple design capable of Mach 3.

Sharply swept, almost delta, wings were mounted at the shoulder of the slab-sided fuselage, at about the midpoint between nose and tail. A conventional set of triangular tail surfaces provided aerodynamic control and stability. A single underslung scoop inlet fed air into the propulsion system, which exhausted through a single nozzle. The cockpit was located in the nose and was provided with a large transparent canopy… not for forward vision, but almost just as a roof, since the canopy was flush with the upper surface of the fuselage. The overall effect was something that might have been doodled by a bored 1950s kid dreaming about rocketships. But the simple configuration belied a complex and ingenious engine.

After considerable competition and study, in July 1951 the AP-57 design was selected by the Air Force for continued development. At the same time, Convair's proposal was also selected for further development. This design was less technologically aggressive than Republic's and went on to become the F-102. Republic continued to refine the AP-57 (officially designated XF-103 in September 1951), with all areas undergoing revision – particularly the cockpit.

Given the high speed expected of the XF-103, a great deal of design work and revision was put into puzzling out how the pilot could see outside in order to fly, with windows that neither created too much drag nor melted due to high aerothermal loads. One option was a small relatively conventional canopy allowing forward vision, with additional drag and poor side and rear vision. Another was a fully recessed cockpit within the cylindrical forward fuselage with only side windows, giving great side vision but no forward vison to speak of. Yet another was a smallish raised housing containing a forward-view periscope. Married to side windows, the periscope – a version of which was flown with considerable success on a Republic F-84G Thunderjet – provided perfectly adequate vision for takeoff and landing. Given the mission of the craft, anything more than that was not seen as strictly necessary. There would be no dogfights for the F-103 – it would engage the enemy at a range of miles, steered by radar.

It was expected that on occasion the F-103's pilot would need to bail out, but ejecting a pilot wearing a standard flight suit would not be an option at Mach 3. Taking a cue from Convair's B-58, the XF-103 was to be equipped with a jettisonable capsule for the pilot. But where the B-58 capsule was essentially a clamshell that enveloped the pilot and his seat, the capsule for the XF-103 pilot was much more extensive. It was to be fitted with deployable stabilizing fins along with an upwards-sliding blast shield that would close off the pilot.

Controls for the aircraft were located within the capsule so he could continue to fly the aircraft even after complete encapsulation. On the ground, the capsule would lower from the aircraft so the pilot could easily get into his seat and then be raised up into the cockpit. It was convenient for normal operations, but neither ejection at low altitude nor runway ejection would be possible.

A constant through the many long years of the XF-103 development was the propulsion system: a hybrid turbojet-ramjet system that offered great promise for Mach 3+ speeds. The XF-103 had a single Wright XJ67-W-1 turbojet – a licence-built Bristol Olympus. It was also fitted with a ramjet engine aft of the turbojet. For high-speed operations valves just inside of the inlet would close off airflow to the turbojet and redirect it to a duct that bypassed the XJ67 and directly fed the ramjet. The XJ67 was temperature-limited to under Mach 3, while the ramjet could operate at up to Mach 4. This composite propulsion system would operate as an afterburner-equipped turbojet from takeoff to below Mach 3, and as a pure ramjet to just above Mach 3. The design of the nozzle changed substantially over time, but for much of the lifespan of the XF-103 programme the nozzle was of square cross-section. This facilitated a 2D variable exit nozzle as well as side-mounted flat speed brakes.

The aircraft structure was the limiting factor in top speed. Aluminium and steel were considered for the primary materials but titanium was determined to be the best choice... it was the lightest and had the best high-temperature properties. This was a bold decision in the early 1950s, given the relative scarcity of titanium and the limited experience American industry had not only in working with it but even just procuring it. In the late 1940s, titanium was only being produced at a rate of a few tons per year in the United States. While that rate increased by several orders of magnitude over the next decade, titanium, its production, manufacture and proper utilization took a long time to become well characterized.

From the outset, the XF-103 was designed for carry Hughes XAAM-2A missiles (soon to be designated the GAR-1 Falcon) and the Hughes MA-1 fire control system. It would have no gun but could manage a mix of missiles. The initial AP-57 design from 1951 would have six Falcons in a bay just behind the cockpit, launching from the upper fuselage; in addition it would carry 36 x 2.75in folding fin unguided rockets in a bay under the fuselage, behind the inlet. The Falcon rockets had a range of six miles or so, but the unguided rockets had a maximum effective range of little more than a mile. They would be launched in a volley to have a faint hope of striking an enemy bomber, doubtless in a head-on aspect with a closing velocity approaching Mach 4. The XF-103 retained this weapons load through at least 1954.

The final interceptor form of the XF-103, dating from January of 1957, still had four Falcon missiles (updated to GAR-3 configuration), but eliminated the short-range unguided rockets. New to the system was the addition of two GAR-X missiles, the design of which began in 1956. These Hughes weapons were larger and more advanced than the GAR-1 Falcons, with folding fins for more compact stowage and 125lb warheads, either conventional high explosive or nuclear.

Work on GAR-X halted in late 1956, but the concept continued to be incorporated into various aircraft proposals and throughout 1957 and 1958 derivative concepts, again called the GAR-X, were produced. These eventually became the GAR-9 then the AIM-47 Falcon. The GAR-X of 1956 vintage gave a range of up to about 25 miles… not great compared to what would soon come, but far superior to that of the folding fin rockets.

While the XF-103 was designed to fly at unprecedented speed and altitude – up to 75,000ft – its range was shockingly low due to the ravenous fuel consumption of the ramjet. Area intercept missions had a radius of only 215 nautical miles and general air defence missions had a radius of only 375 nautical miles; time spent at Mach 3 under pure ramjet power was only about five minutes. Range could be extended with the use of underwing drop tanks, but even with those ferry range was only 1343 nautical miles. For a point defence interceptor this was seen as adequate, but it would certainly limit employment.

In order to provide coverage of the entire continental United States, operational F-103s would have needed to be scattered widely, likely with a great many spread across the Canadian wilderness. Continental US defence missions would be just about all the F-103 could manage since ferry range was insufficient to cross the Atlantic. Overseas deployment would require transport by either ship or cargo plane. Inflight refuelling was not a part of the design and given the pilot's lack of clear vision this was likely for the best.

The XF-103 was in development for a surprising length of time, from 1951 to 1957. While this would be considered a trivial span of time today (barely covering the time Congress would need to debate what the colour of the aircraft's wheels should be), in the 1950s this was unacceptable. When the XF-103 began life it was a futuristic advanced technology showcase; by 1955 it was obsolete and, what's more, it was becoming clear that it would probably never be able to fulfill its original promise.

In 1955 the Air Force issued a new General Operating Requirement (GOR 114) for a Long Range Interceptor, Experimental (LRI-X), a new interceptor mandated to have two crew and two engines. This excluded the XF-103 but the programme was not instantly killed; it dragged on for a few more years, and in early 1957 it was turned into merely a research vehicle to test avionics and radar for the LRI-X aircraft to come. In August of 1957 the XF-103 programme was finally officially cancelled.

The XF-103 is typically depicted in a bare metal finish. This is true not only of modern artwork, but the art, models and mockups of the time also showed the aircraft with a shiny bare titanium skin. While this is certainly pretty, it seems that Republic engineers had intended to apply a blue (presumably dark blue) high-emissivity coating to the aircraft. This would serve the same function on the XF-103 as the near-black coating on the A-12: to radiate heat away from the titanium skin and help cool the structure.

The XF-103 illustrated here is how the design appeared during an inspection in January 1957 – its final interceptor configuration before its conversion, in July 1957, to an unarmed research-only layout.

Long Range Interceptor – NAA XF-108

The October 1955 release of GOR 114 was official recognition that the Republic XF-103 was not going to see service as an operational interceptor. Recent technological advances had shown the Air Force that

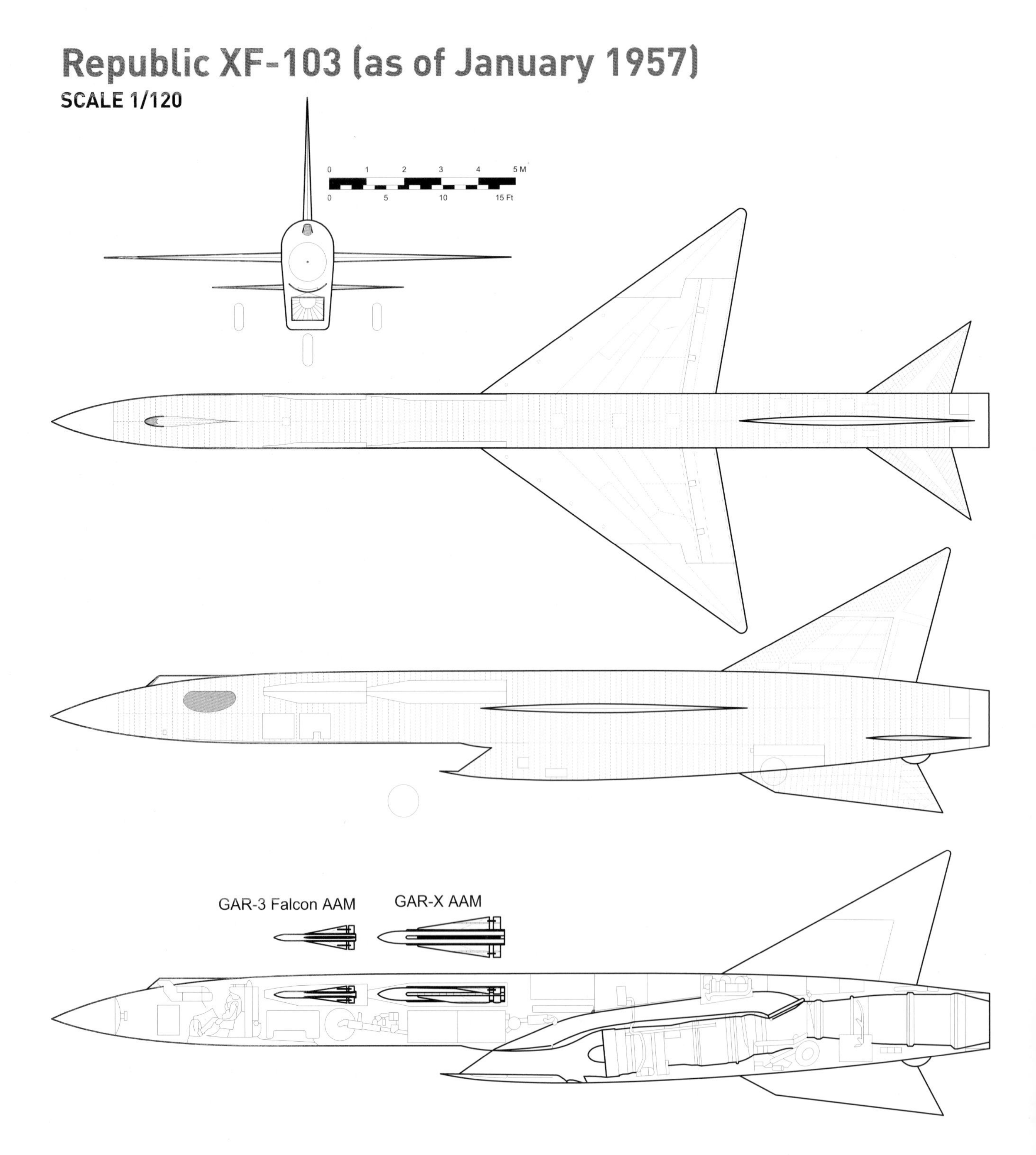

a more conventional sort of turbojet could reliably shove a capable interceptor to high Mach and keep it there. A twin-engine, two-man craft would be much more reliable and capable. Nevertheless the requirements continued to push the technological envelope of the time.

Beyond the additional crewman and engine, GOR 114 also added considerable range. Gone was the XF-103's role as a point defender; the new aircraft, the Long Range Interceptor – Experimental (LRI-X) would have to have a range of 1000 miles. This put it in a league far beyond what the fuel-thirsty XF-103 could attain. In early 1956 Lockheed, Northrop and North American turned in their designs; the Lockheed and Northrop designs were impressive, but North American's outstripped them both and was selected.

The NA-236 design was a large delta with canards and Mach 3 capability. The fuselage and engine layout was similar to that of the contemporary North American A3J (later A-5) Vigilante, with a single two-dimensional inlet on either side of the fuselage. But unlike the Vigilante, the NA-236's engines were closely spaced, side by side at the extreme rear of the fuselage. The aircraft had a shoulder mounted delta wing, relatively close-coupled high-mounted delta canards and a single large vertical stabilizer. It also had smaller vertical stabilizers mounted mid-span that extended both above and below the wing as well as short outward-canted dorsal fins on either side of the rear fuselage, all meant to counter a stability issue that would remain with the aircraft till the end. The main structure was to be titanium with a steel honeycomb skin.

In typical bureaucratic fashion, the USAF cancelled the LRI-X programme in May 1956 due to infighting about what kind of aircraft was needed... interceptor vs fighter, short range vs medium range vs long range. The following April, the Air Force decided it had got it right the first time and reinstated the long range interceptor programme, adding in a requirement to reach Mach 2.5. Further, in June 1957 the Air Force decided that North American's was still the best design and elected to provide a contract for two prototypes, officially designated F-108. Meanwhile, North American had revised the design, producing the NA-257, though the new design was visually little changed from the NA-236.

Work began in earnest and the design began to rapidly evolve. The canards disappeared (last seen in May 1958), leaving a much simpler-looking delta configuration. The configuration from September 1958 had simple delta wings, slightly clipped at the tips, with a single large dorsal vertical stabilizer and two ventral stabilizers under the wings at about mid-span.

Something that remained constant throughout the F-108's brief development life was the powerplant: two General Electric J93 engines. These engines were designed for the B-70 bomber, six of them being able to push that relatively vast aircraft past Mach 3. Two would do the same for the F-108. Normal thrust of the J93-GE-3AR was projected as 18,400lb, rising to 20,900lb at military thrust and 29,300lb max thrust with afterburner. The J93 burned a new fuel, JP-6. It was similar to JP-5, a kerosene-based fuel used by the Navy due to its low volatility, making it less likely to catch fire on an aircraft carrier. JP-6 was created in 1956 for the J93 and the B-70 bomber and was chemically modified for a lower freezing point and improved thermal oxidative stability.

For landing at Arctic air bases, which could be expected to be ice covered, the J93s were fitted with thrust reversers. Two clamshell doors, normally stored one above and one below the exhaust nozzle, would translate aft and then rotate to close off the exhaust and redirect it forward and up, and forward and down. This was expected to provide better braking performance than a parachute system, as well as being easier to deal with after each flight. The thrust reversers were important enough to have full scale functional (to the extent that they moved) mockups built and displayed.

The B-70, also a North American Aviation product, was a selling point for the F-108: the engines, materials, aerodynamics and design practices perfected for the B-70 could be applied to the F-108. But where the B-70 carried bombs or surface-attack missiles, the F-108 would carry air-to-air missiles – specifically AIM-47 Falcons. The F-108, somewhat perversely, carried fewer missiles than the less capable F-103, with three AIM-47s on a rotary launcher in a single bay between the intakes, launched through a door in the underside of the fuselage.

For target acquisition and tracking, the F-108 was fitted with an AN/ASG-18 fire control system with a 40in diameter radar dish. This could spot a medium bomber at a range of 100 nautical miles. In addition the F-108 had two infrared search and track sensors, one mounted in each wing leading edge just outboard of the intakes. These could spot a subsonic enemy bomber, when seen from the rear, at a range of about 35 miles, dropping to about 10 when seen from the front. A Mach 3 Soviet bomber, development of which the US Air Force considered likely, would emit far more infrared both from its engines and due to aerothermal heating; the F-108's IRST system should be able to spot such an aircraft more than 76 miles away. However, 1950s computer technology limitations meant the system could only track one target at a time.

The aircraft's design as of October 1958, was approximately its final form. The earlier simple delta wing had been modified into a cranked delta with lower-sweep wingtips. The wings were given a few degrees of dihedral except for the wingtips which had a measure of anhedral, but retained the mid-span ventral stabilizers. In addition, the fuselage-mounted ventral stabilizers were greatly increased in area; they were given hinges to fold up for takeoff and landing. Directional stability at high speed had proven a difficult problem to crack. The vertical stabilizer was all-moving rather than having a separate rudder; the trailing edge of the wing was divided up into a number of separate control surfaces much like the 'finger' elevons of the XB-70.

Also similar to the XB-70 were the separate escape capsules for the crew. Similar in concept to, but simpler than, the escape capsule planned for the XF-103, the XF-108 system would encapsulate the crew in the event of an emergency. If needed, they would then eject and protect the crew down to a parachute-lowered landing; or the crew could stay in the aircraft and control it to a limited degree.

The crew compartments were meant to be pressurized and air conditioned shirtsleeve environments, meaning the crew would not need complex protective flight suits. While this would make flight reasonably comfortable, it would also mean that pressurized escape capsules capable of keeping the crew alive above 70,000ft were a must. With this design iteration, range and duration could finally be extended via in-flight refuelling, using a receptacle on the upper fuselage behind the crew compartment.

At that time, the initial operational capability of the F-108 was estimated to be mid-1963. A mockup was built and reviewed in January of 1959, and in May of that year the XF-108 was officially dubbed the 'Rapier'. But the programme's time was almost up.

Where the XF-103 had failed because it took too long to develop an aircraft that would not have met the needs of the time, the XF-108 was cancelled in large part because it was surplus to actual requirements. In the 1950s there was a substantial fear of the Soviets producing wave after wave of bombers, blotting out the Sun with long range aircraft capable of using canned sunshine to erase American cities at will. But by the late 1950s, the CIA was discovering that those fears were greatly overblown: the Soviets did not have anything at all like the bomber force – or the capability of producing one – that the West feared.

Coupled with the arrival of intercontinental ballistic missiles (a threat that manned interceptors could do little to nothing about) and the development of capable, nuclear-tipped surface-to-air missiles, it became clear that the XF-108 programme, while technically on track, was simply not needed. Given the projected high cost of development of an aircraft that the Air Force no longer felt a burning need for, in late September 1959 the Air Force cancelled the XF-108 contract.

Convair B/J-58

The Long Range Interceptor programme initiated in 1955 resulted in a number of proposals, even well after the North American F-108 began development. One worthy of mention was the Convair B/J-58 of 1959, a modified derivative of the proposed B-58C bomber. In the B-58C and B/J-58, the engines were reduced to two, upgraded to afterburning Pratt & Whitney J58s. As originally projected for the B/J-58, the J58s provided more than double the thrust of the J79, so slashing the number of engines in half still resulted in more thrust, at lower weight and drag.

The crew of the B/J-58 was reduced to two from the B-58A's three (filling that empty space with a new fuel cell) and small vertical stabilizers were added near the wingtips. The nose was greatly increased in size to accommodate the 40in radar dish of the three kilowatt AN/ASG-18 fire control system. The fuselage was stretched by 5ft aft of the crew compartments to provide yet more fuel volume. Consequently the vertical stabilizer was increased in area to accommodate the need to offset the yaw instability imparted by the larger nose. Infrared search and track sensors were built into each wing leading edge near the roots; the wing leading edges were themselves extended, increasing wing area and leading edge sweepback.

The M61A1 Vulcan gun was removed from the tailcone (which was itself recontoured) as an interceptor would be unlikely to need one. And importantly, the belly pod essential to the B-58 bomber was replaced with an all-new one containing three GAR-9 missiles. The pod would have a set of clamshell doors for each missile, similar to the missile bay doors on the F-106.

The B/J-58 LRI was less optimized for the interceptor role than the F-108, but it had the perceived advantage of being based on an established airframe. Additionally, important parts of the B/J-58 did fly in a sense: in 1959 Convair modified a single B-58 to carry the 40in radar dish – complete with larger fibreglass nose – and a missile pod. In this case B-58 serial number 55-665 (nicknamed 'Snoopy' due to the large drooped nose) was meant not to be an actual interceptor but a test article for the relevant systems. Snoopy was not build to aid the B/J-58 proposal, but the North American XF-108. The pod contained and could launch a single GAR-9 missile. Snoopy included the infrared search and track sensors, but located them alongside the nose rather than in the wing roots. In 1962, Snoopy began a series of test launches of the GAR-9, showing successes at intercepting target drone aircraft. It last launched a GAR-9 – by then redesignated the AIM-47 – in 1964.

North American F-108

SCALE 1/144

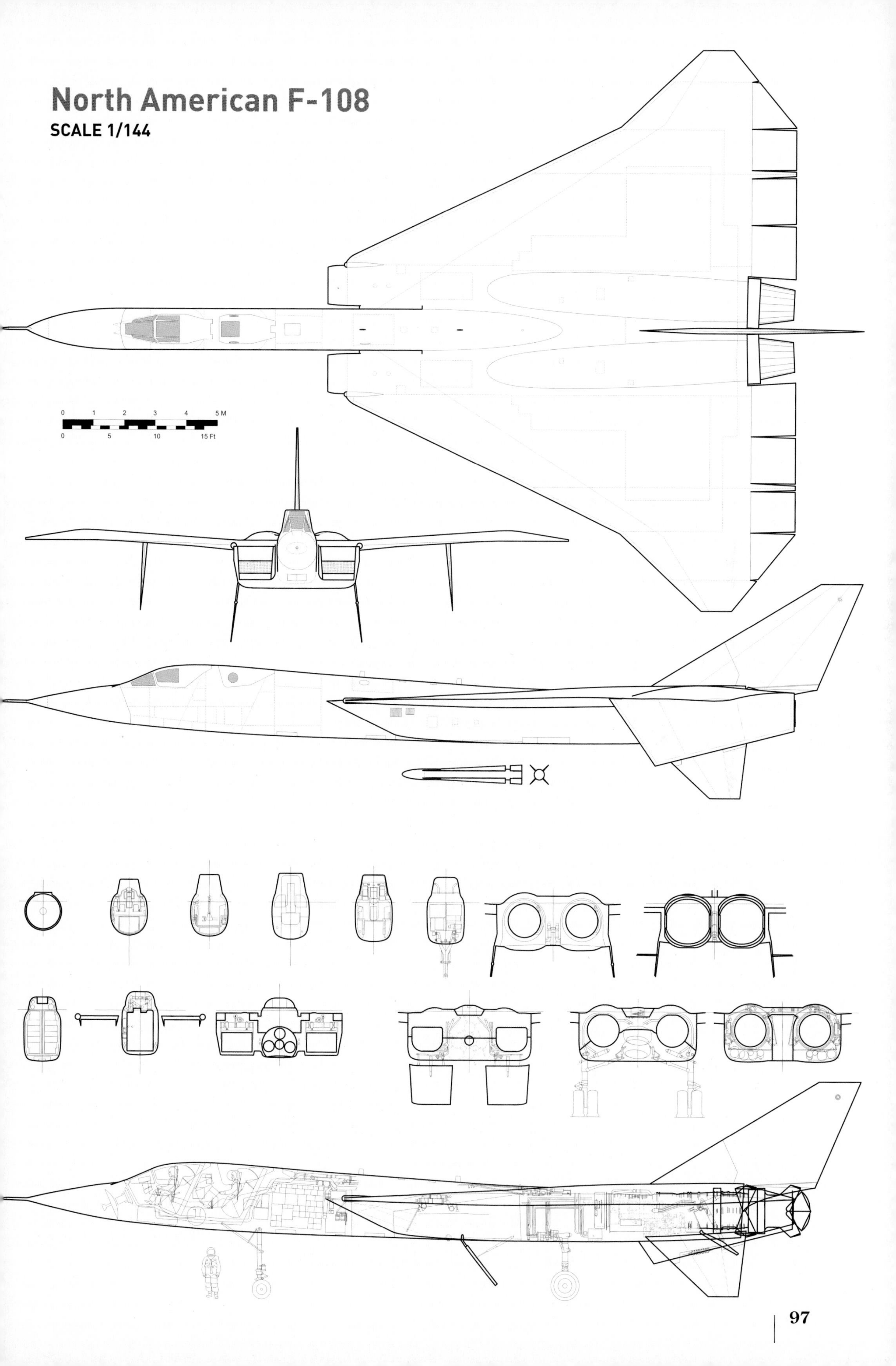

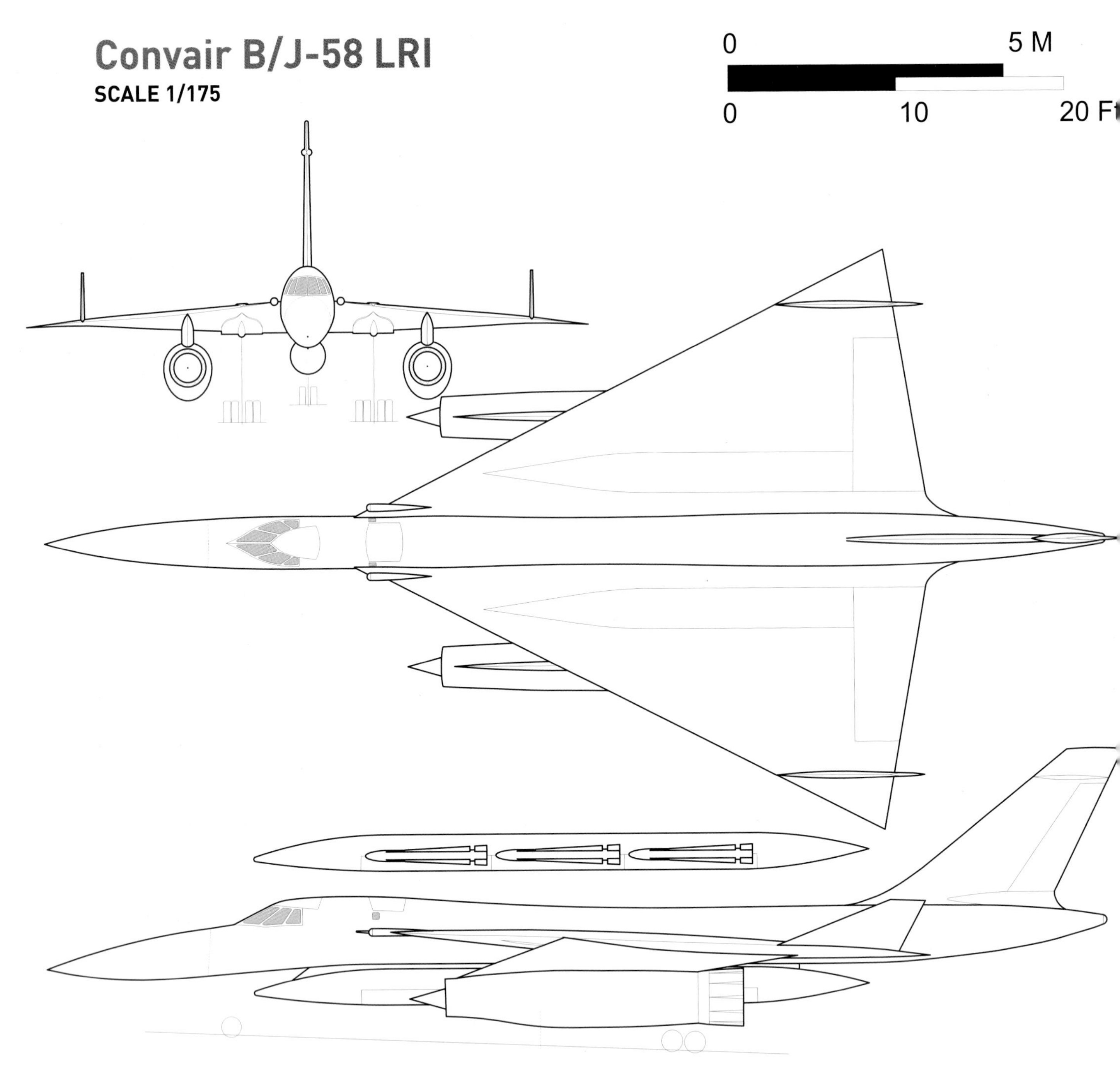

Lockheed A-12 Derived Interceptor

The North American F-108 Rapier was dead. But the weapon it was meant to carry – the GAR-9, backed up by the AN/ASG-18 fire control system – was still alive. The Air Force realized that the missile itself was potentially the answer to the issue of blowing Soviet aircraft out of the sky at long range, even if the Rapier was not the aircraft to carry it. So development of the GAR-9 continued, with launches being carried out using Snoopy. And the American aviation industry wanted to produce an aircraft to carry and launch the GAR-9.

This of course included Lockheed. And by 1960 the company already had an aircraft almost ready to go: the A-12 reconnaissance plane. It was as fast as the F-108 was going to be, with the added bonus that at that time development seemed almost certain to result in a flying aircraft. It was probably inevitable that Kelly Johnson would suggest using the A-12 as an interceptor.

The early design history of the interceptor variant of the A-12 is still somewhat obscure. What is known is that an initial proposal to the Air Force was made for what was dubbed the 'AF-12'. This would have taken an incomplete A-12 from the assembly line and modified it for the interceptor role by adding missile bays (for folding fin versions of the GAR-9), a search and track radar system and a second crew station. No diagrams of this seem to have come to light, only the barest description.

It can be speculated that the aircraft would have most closely resembled the M-21, with its very slightly protruding second crewman station. This configuration was accepted and the Air Force was sufficiently pleased that in October of 1960 a contract to fully develop the AF-12 was signed, specifying that the seventh through ninth A-12s in the production line would be modified into the interceptor. The programme was codenamed 'KEDLOCK'. This code name – kedlock being an obscure name for a mustard plant – would have certainly proven unenlightening to any Soviet spies who came across it.

Work continued to a mockup review in May 1961. Wind tunnel testing resulted in the addition of three ventral fins in June of 1961... two smallish fins fixed to the underside of the nacelles, one hydraulically actuated large fin under the extreme rear of the fuselage. This became necessary because some time prior to that the nose had grown larger to accommodate the 40in radar dish of the AN/ASG-18 fire control system. The larger nose made the basic design unstable at high speed, necessitating the additional stabilizer area. The inclusion of a large diameter radar meant that the cockpit needed to be moved upwards in order for the pilot to see over the nose. This gave the resulting aircraft a somewhat 'humped' appearance.

Lockheed AF-12 (1961)

The earliest known diagrams of the AF-12 date from 1961 and depict the aircraft as it almost ended up being built. This AF-12 design had all the features that would eventually be built, but with some slight yet notable detail differences.

One oddity was with the central folding ventral fin. The fin did not deploy straight down while in flight, but some 22° offset to port, perpendicular to the fuselage at that position. Why it was not mounted on the centreline, and consequently why it didn't deploy purely vertically, is unclear. Also, the chines extended a short distance forward onto the new nose cone, with a rounded curvature in the plan view.

Lockheed YF-12A (1962)

The AF-12 configuration rapidly evolved. The folding ventral fin soon found itself on the centreline; the chines were cut back further. The nose of the AF-12, once the AN/ASG-18 was added, became fundamentally different to that of the A-12. Gone were the graceful chines all the way to the pitot tube; instead there was a great ogival cone, featureless from the pitot tube back to the cockpit. The nose was now a single piece of composite plastic, transparent to radar and resistant to both the high thermal and aerodynamic loads to be expected from Mach 3+ flight.

The chines were unceremoniously chopped off and given an unswept sharp leading edge. At the corners they were fitted with infrared search and track units such as the F-108 would have carried. These were positioned much more like those on the Snoopy test vehicle than they were on the F-108 design.

As development and construction of the A-12 continued, the AF-12 continued right alongside it – concurrent, but with added layers of security classification complexity thrown in. Construction of the first AF-12 began by August, 1962, at the Lockheed facility in Burbank, California. In September 1962, the 'Tri-service aircraft designation system' was instituted by the Department of Defense, unifying the way that Navy, Army and Air Force aircraft were designated under the then-current Air Force system. As a result, the AF-12 became, rather conveniently, the F-12; the prototypes then under development were given the Y-for-prototype prefix, becoming YF-12As.

All three YF-12A airframes were well along by spring of 1963. When completed they were trucked to Area 51 in Nevada for flight testing. The first flight was on August 7, 1963; at the same time, assembly of the second airframe was well under way. The second airframe first flew November 23, 1963, the third on March 13, 1964.

The YF-12A was the first of its A-12 derived stablemates to be revealed to the public. On February 29, 1964, in his first nationally televised news conference, President Johnson announced the existence of the 'A-11' aircraft. Included in the disclosure were two photos of the YF-12A, one in flight and one sitting on the ground. The photos showed the aircraft from much the same angle... more or less straight side-on from the aircraft's starboard side. Aviation periodicals around the world ran with the story, excited over not just the announcement of a new aircraft but a *remarkable* new aircraft capable of more than Mach 3.

With only one view of the craft, artists, illustrators and draftsmen did their best to produce three-view diagrams and even cutaway illustrations; and without exception the plan views were quite inaccurate. Some were valiant efforts; some were frankly horrible. But for a little while at least, the public was uncertain about just what exactly the 'A-11' looked like.

The 'A-11' designation was selected by Kelly Johnson as a bit of deception. The A-11 design was of course notably unlike the YF-12, being purely a reconnaissance aircraft having no concessions to radar cross section reduction. The assumption seems to have been that had the Soviets penetrated Lockheed and CIA security and obtained information on the programme, the A-11, one of the more heavily studied of the precursors, may well have been the design the

Lockheed AF-12

SCALE 1/120

General Data
- Crew: 2
- Span: 59ft
- Wing Area: 1795sq ft
- Length: 101.66ft
- Powerplant: 2 J58 w/afterburners
- Gross Weight: 124,000lb
- Landing Weight: 68,000lb
- Armament: 3 GAM-9

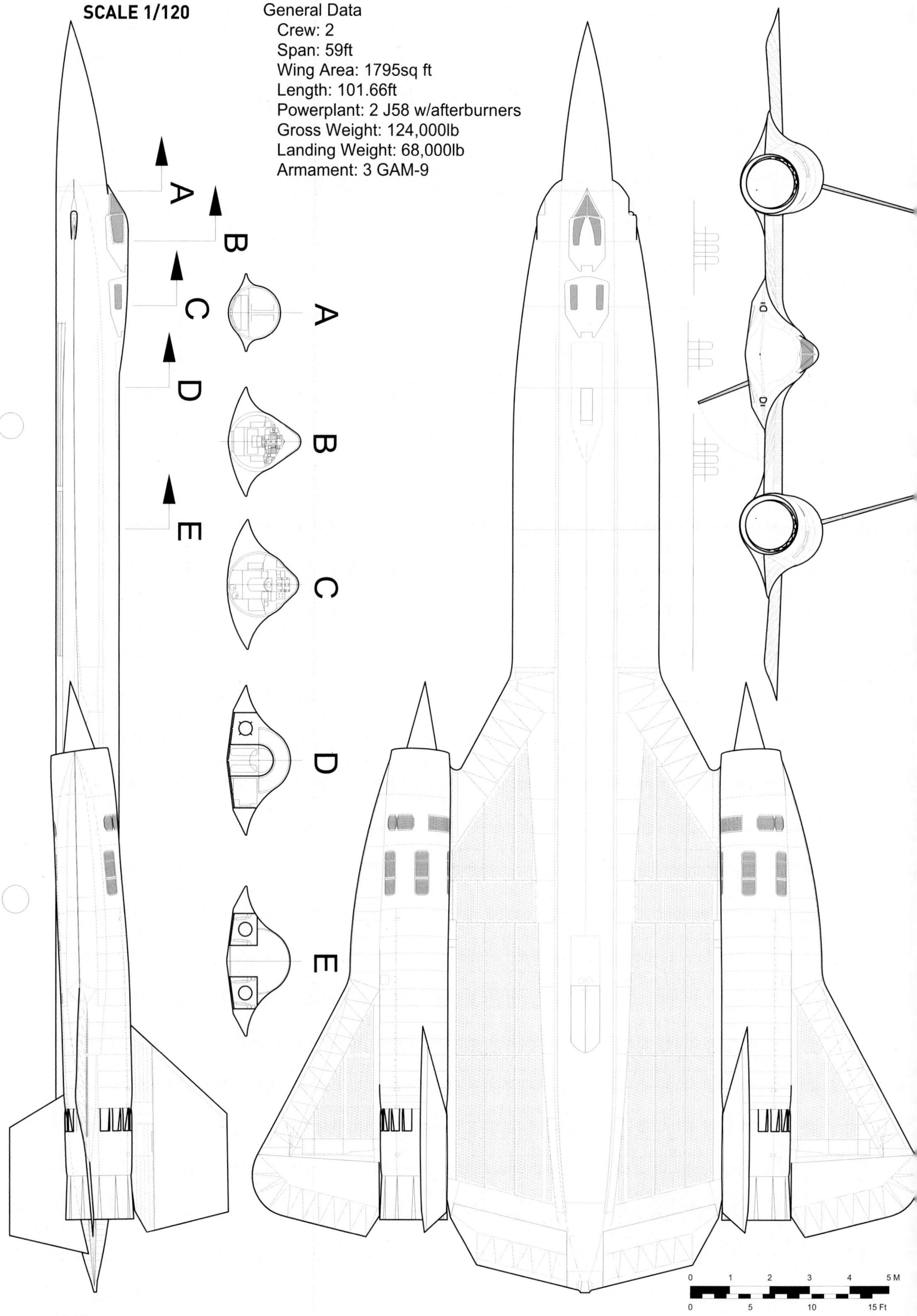

A B C D E

A

B

C

D

E

0 1 2 3 4 5 M

0 5 10 15 Ft

Soviets had the most information on. So by referring to the new aircraft as the 'A-11', some pressure might be taken off the actual A-12.

The YF-12A had recently begun flight testing, and the A-12 was expanding its flight testing programme with the hope of operational flights in the near future. Consequently, it was decided to get information out about the YF-12 – even if mislabelled – so that if anyone happened to spot one they would assume that it was the much less sensitive 'A-11'. The A-12 and the YF-12A were sufficiently similar in configuration that laymen who happened to spot an A-12 would assume that they had seen the 'A-11'. Curiously, the F-12 was for a time referred to, at least within the CIA, as the 'X-22'; at the same time, the A-12 was referred to as the 'X-21', with the possibility of being revealed to the public as the 'R-X' or 'RX-12'. The A-12 itself would remain unknown to the public until revealed by the CIA in 1982.

The aft fuselage of the YF-12A was, apart from the ventral fins and hardware associated with folding the central fin, basically the aft fuselage of the A-12. The forward fuselage, however, was notably different. Along with the larger nose and raised cockpit, the underside of the forward fuselage was extended outwards somewhat, changing the outer mold line.

The cross-section of the A-12 had featured straight lines from the end of the chine, tangent to the circular fuselage. But the YF-12A pushed the centrepoint of that connecting line outwards, increasing internal volume. This provided enough thickness for missile bays that could fit the GAR-9. Four such bays were provided, two on either side, the bays in tandem. The bays had doors split lengthwise down the middle, hinged on the outer edges. Although four bays were created, only three missiles were carried. The forward right-hand bay contained the ASG-18 fire control system computers, with a single-piece door that was designed to drop downwards to lower the computer equipment for servicing.

The GAR-9 missile and associated radar and fire control systems of course predated the YF-12A. Nevertheless, learning how to properly employ the weapons – in particular high-speed ejection from the aircraft – proved to be anything but easy. The first in-flight ejection of an unpowered but accurately configured and weighted test round occurred on April 16, 1964. And while the test 'missile' did successfully separate from the aircraft, it did so at an unfortunate nose-up angle. Had it fired its rocket motor it would have struck the YF-12A in the vicinity of the cockpit, clearly an undesirable outcome. Lockheed redesigned the mechanism, resulting in a successful system.

Through 1966, 13 missile launches were conducted from the YF-12A. None were armed with live warheads, but they proved capable of passing close enough to the targets that had they been armed the targets would have been destroyed. In March of 1965 the YF-12A scored its first kill, firing a GAR-9 at a target 36.2 miles away. In September of 1965, the first YF-12A launched a GAR-9 from an altitude of 75,000ft and a speed of Mach 3.26 (the intended and expected launch condition for an operational F-12), successfully intercepting a target 32.2 miles away, missing by little more than 6ft.

The AN/ASG-18 coupled with the GAR-9 was an incredibly capable system for the time, able to see and attack targets not only at a range of 100 miles, but also down to an altitude of 500ft. The radar could distinguish an aircraft from the background, something that other, smaller systems could not accomplish. The YF-12A could in theory completely dominate the sky.

Lockheed AF-112D/F-12B (1964)

The YF-12A aircraft were always meant to be prototypes, not operational aircraft. The hoped-for F-12B would be a more optimized design with a number of notable revisions to the design. But as with the early history of the AF-12, the specifics of the F-12B are muddled, available only as scraps. One design that seems likely to be representative of the F-12B is the 'AF-112D' from 1964. Most obvious of the changes from the YF-12A was the return to a fully chined nose and the lowering of the cockpit. The appearance is much more like that of the SR-71, through the cockpit also seems to have been moved forward by about half a foot. The underside of the forward fuselage was also flattened, much like the SR-71, and the tail of the fuselage is extended like the SR-71.

The underside of the rear fuselage also showed some differences. The central large stabilizing fin was deleted (likely due to the smaller side profile of the nose), leaving the smaller fins on the underside of the nacelles. Where the central fin would have been there was now a low fairing serving a singular purpose: protecting an arrestor hook. This appears to have been essentially identical to the arrestor hook used on the A-12CB, though it is unlikely that the F-12B would ever land on a carrier. Instead the likelihood is that carrier-like cables might have been planned for use at smaller air bases, allowing the F-12B to land where it might not otherwise have been able to.

The F-12B, like the YF-12A before it, included a load of three GAR-9 missiles, with the fire control system contained within the fourth missile bay. The bays were necessarily smaller since the underside was compressed compared to the YF-12A; as a result the missiles only just fit. But the YF-12A used ejectors

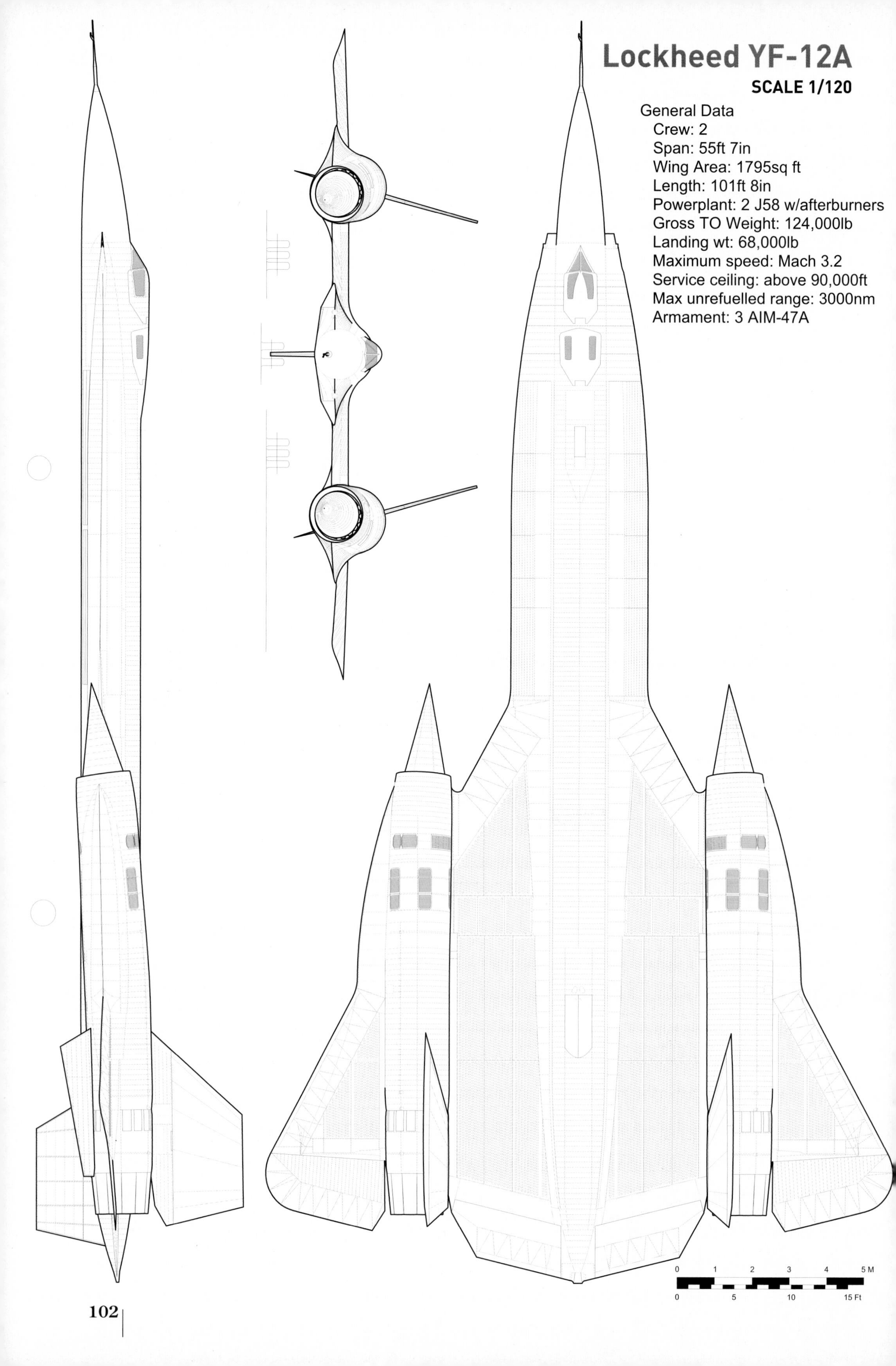
Lockheed YF-12A
SCALE 1/120
General Data
Crew: 2
Span: 55ft 7in
Wing Area: 1795sq ft
Length: 101ft 8in
Powerplant: 2 J58 w/afterburners
Gross TO Weight: 124,000lb
Landing wt: 68,000lb
Maximum speed: Mach 3.2
Service ceiling: above 90,000ft
Max unrefuelled range: 3000nm
Armament: 3 AIM-47A
0 1 2 3 4 5 M
0 5 10 15 Ft

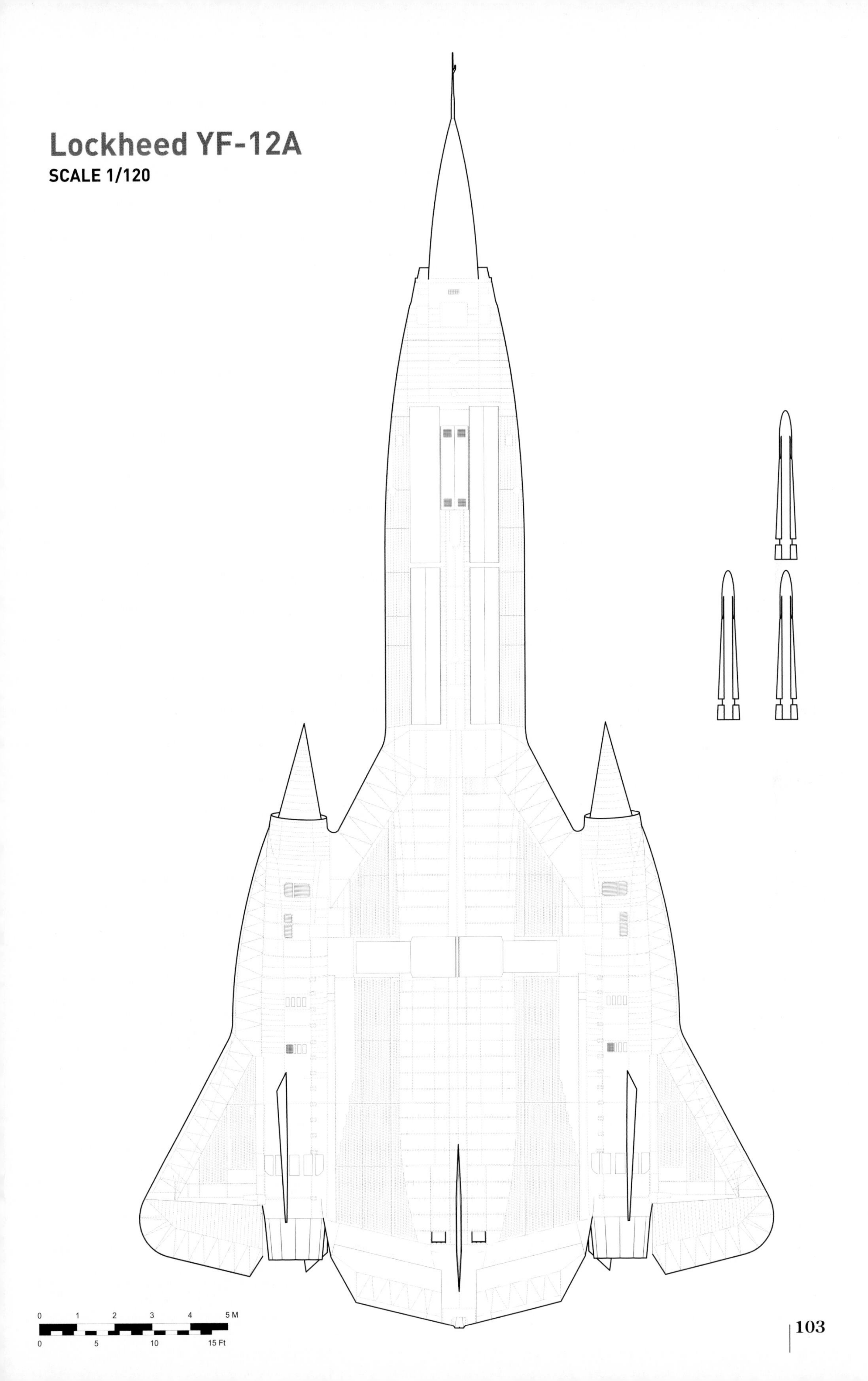
Lockheed YF-12A
SCALE 1/120
0
1
2
3
4
5 M
0
5
10
15 Ft

to toss the missiles from the bays, while the F-12B was to use a deployable trapeze launch rails. Notably, the rails would deploy the missiles in a substantially nose-down attitude.

As of early 1967, Lockheed had proposed to build 100 F-12Bs. These would be spread out to different Air Force bases, with three squadrons of 16 aircraft each sent to bases in the northeast of the US, and three squadrons of 16 aircraft each sent to bases on the west coast. Unfortunately, Secretary of Defense McNamara did not support the development of the F-12B. In late December of 1967, orders came down from the Air Force to cancel all air defence programmes, ending the hope of a production F-12B.

Lockheed FB-12 (circa 1967)

One of the more highly modified of the Blackbird studies was what is apparently designated the 'FB-12'. Known solely from a series of Skunk Works diagrams with the data blocks sadly unavailable, the FB-12 did something unique to the venerable configuration: it bent the forward fuselage. Ahead of the FS 715 field joint, the fuselage was bent upwards by about 2.1 degrees; the nose and the cockpit were then bent downwards by a roughly equivalent amount. This resulted in a configuration that looks like a snake ready to strike, with the underside of the forward fuselage – and the four missile bays similar to those on the YF-12A – visible from the front.

Why this was done is unclear, though the modifications would have been major enough that undoubtedly some purpose called for it. Chances are wind tunnel testing had shown that a raised forward fuselage would improve the aerodynamics of the vehicle at cruise conditions. Available official diagrams of the FB-12 do not depict the rear fuselage, so it is unclear if the design would have included the central ventral fin. However, NASA testing showed that it was not strictly needed, so in these reconstruction diagrams it is not shown.

As the designation 'FB-12' indicates, the design was to be used as both a fighter and a bomber, sometimes at the same time. At least six distinct weapons loads were drawn and it is possible that more were produced. The weapons loads came with some differences in radar units to be carried; unlike the YF-12A, no infra-red seekers are in evidence. All target acquisition and aiming seems to have been via radar. The nose had the large diameter of the YF-12A to fit large radar dishes, but it was fully faired. Notably, the chine had a larger diameter along the leading edge down the whole length of the fuselage. It's unclear exactly how that would blend into the relatively sharp leading edge of the wing.

The modified forward fuselage would have been more than purely cosmetic. Doubtless components would have been used from the SR-71/F-12, but the overall structure would have had to be almost entirely new.

The first weapons load illustrated was four Boeing AGM-69 SRAM missiles. This would be a pure bomber, with the Short Range Attack Missiles occupying the missile bays. The SRAM missile was the standard US Air Force tactical nuclear strike missile from 1972 until 1993. It carried a W69 warhead with a variable yield of either 17 kilotons (approximating the power of the Fat Man and Little Boy bombs) as a pure fission device, or 210 kilotons using tritium boosting. As the name suggests, the SRAM was capable of only short ranges, generally about 110 nautical miles, reaching a top speed of Mach 3. But the FB-12 would have doubtless launched the missiles at high altitude and triple-sonic speeds, notably improving their range. Even so, the FB-12 would need to penetrate well into enemy territory.

There were two radar options for this weapons load. The primary radar was the Westinghouse AWG-10, used on the F-4J Phantom for guidance of AIM-7 Sparrow missiles. An alternate system was a side-looking radar for scanning the terrain below, doubtless for precise SRAM aiming.

The existence of SRAM missiles in the diagram helps to put an earliest date on the FB-12 configuration: Boeing did not win the AGM-69 contract until 1966. Lockheed designers doubtless would not have included Boeing missiles in their layouts while Lockheed was still competing for the programme, even if they had good diagrams of their competitor's design.

The second weapons load illustrated was a concept apparently designated the FB-12-4. This had a load of two SRAM missiles for surface attack and two 'AIM-7E/F' missiles for defence. These would have been folding-fin versions of the AIM-7 Sparrow air-to-air missile. The Sparrow used semi-active radar guidance, and if the FB-12 was indeed designed in 1966 or 1967, was already racking up a dismal performance record in engagements over Vietnam. Given that the FB-12 would doubtless have cruised at an altitude few enemy aircraft could hope to match, it's unclear just what the AIM-7s would be used for, unless it was planned for the FB-12 to use the two air-to-air missiles to help clear the path to the target.

This vehicle would have baselined the AWG-10 radar system. However, an alternate radar system was suggested using either the General Electric APQ-114 Ku-band attack radar (employed on the F-111A) or the Rockwell Autonetics APQ-130 attack radar (used on the F-111D). The inclusion of Sparrow missiles may also help narrow down the date of these designs, as Kelly Johnson is known to have discussed conversion of both A-12 and SR-71 aircraft to fighters armed with Sparrows in November 1967.

Lockheed AF-112D

SCALE 1/120

General Data
- Crew: 2
- Span: 59ft
- Wing Area: 1795sq ft
- Length: 1246.67in
- Powerplant: 2 J58 w/afterburners
- armament: 3 AIM-47

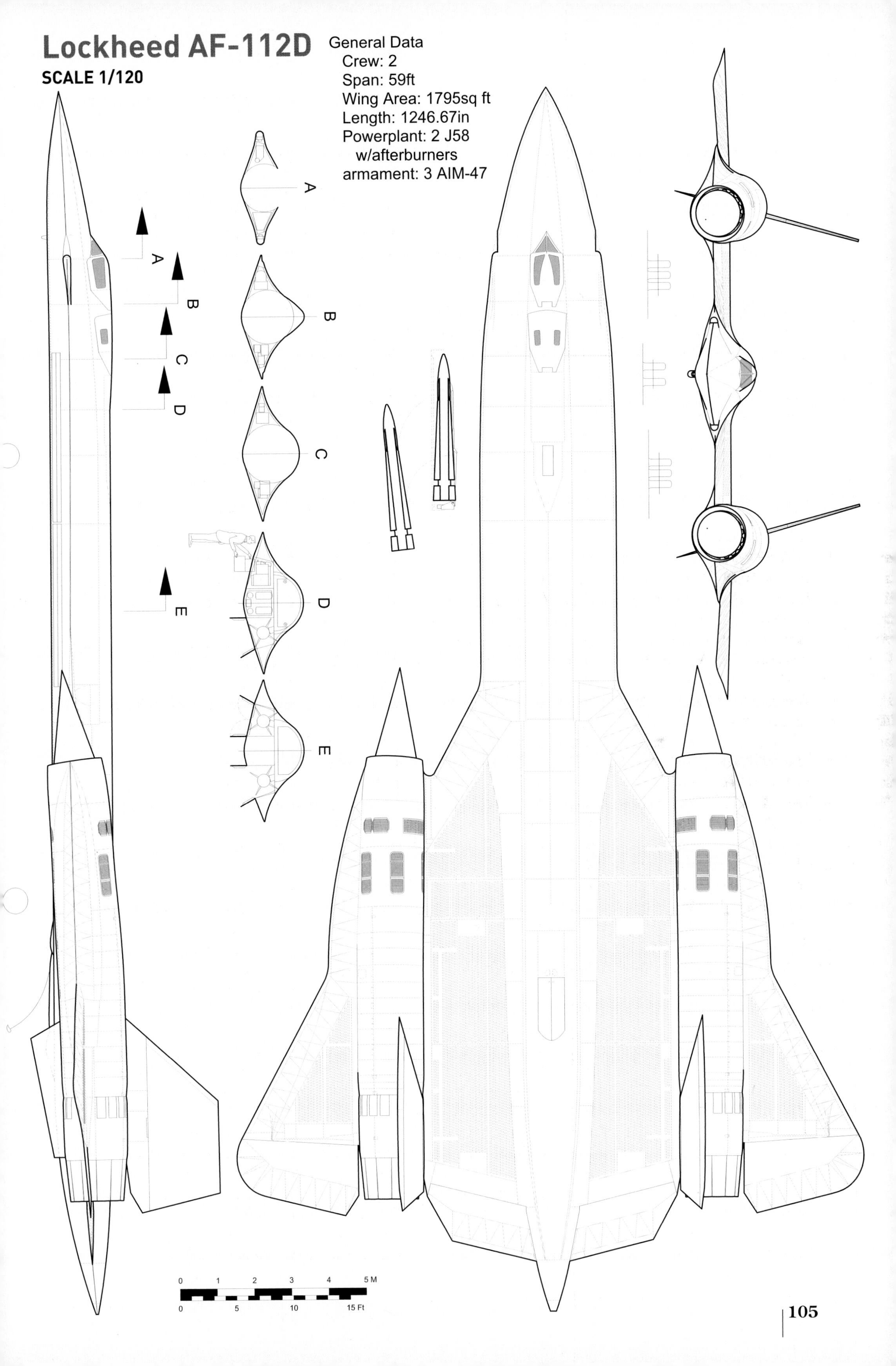

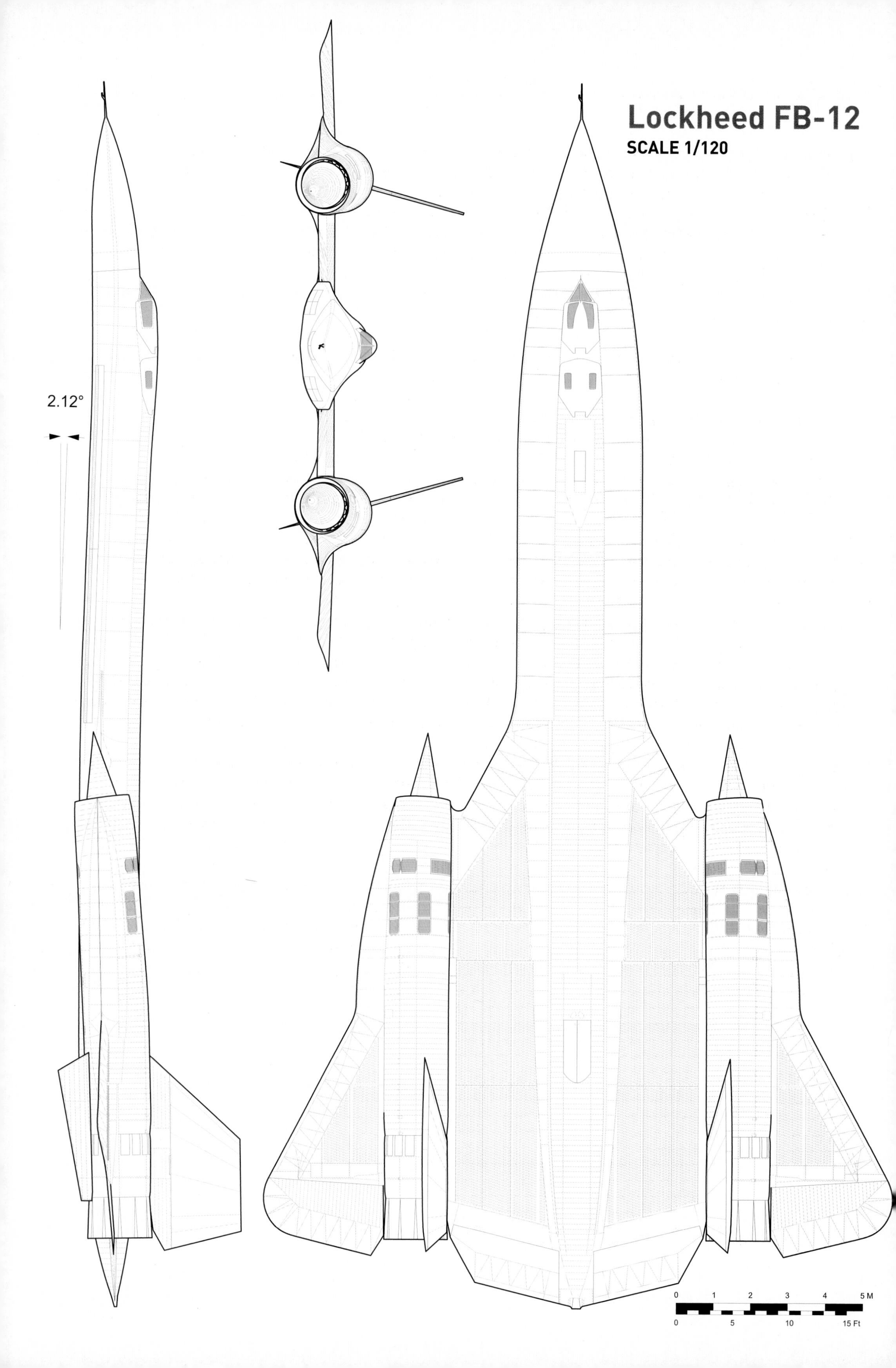

Lockheed FB-12
SCALE 1/120
2.12°
0
1
2
3
4
5 M
0
5
10
15 Ft

The third load was four AIM-7E/F missiles. This would be a dedicated interceptor armed with four medium range missiles with medium-sized conventional warheads; this seems like a bit of a waste of the aircraft's potential.

The fourth load was four AIM-47B Falcon air-to-air missiles. Similar to the YF-12A, but with a folding-fin version of the missile for more compact stowage. It's unknown whether the AIM-47B would be equipped with a nuclear warhead, as the AIM-47 was originally intended to have, or the 100lb conventional explosive warhead the AIM-47 ended up with. In either case, the FB-12 with AIM-47Bs would have had much greater ability to reach and destroy enemy aircraft than the Sparrow-armed versions, and one more missile than either the YF-12A or the F-108 would have had. To find targets, this version of the FB-12 would be equipped with the Hughes AWG-9 planar array radar, originally designed for the F-111B and its load of AIM-54 Phoenix missiles. The AWG-9 entered service in the 1970s on the F-14 Tomcat.

The fifth and sixth configurations were probably the most interesting and certainly the most inexplicable. Each had three air-to-air missiles and one General Electric M61A1 Vulcan Gatling gun. What the FB-12 would do with a fixed gun is murky at best, as none of the Blackbird designs was anything remotely like a 'dogfighter'. The gun was carried in the port forward missile bay and was fixed at a slightly downward angle. It is beyond unlikely that the FB-12 would use its gun to strafe ground targets; on the other hand, it is very likely that it would be expected to fly substantially higher than most aircraft it might be targeting. Thus shooting slightly downwards makes sense, even if the difference in engagement altitudes might well have been extreme.

The two missile loads were envisioned were, firstly, three AIM-7E/F medium range air-to-air missiles, the same folding-fin version previously described. This layout included the AWG-10 radar for target acquisition and missile guidance. And secondly, three AIM-47B long range air-to-air missiles, using the AWG-9 phased array radar. This is even more difficult to understand than the version with AIM-7E/F, since the Falcon missiles would allow engagements with aircraft 100 miles away, while the gun would require effectively point-blank range.

The YF-12A did not end its existence as many unsuccessful development programmes have – being chopped up for scrap or summarily shuffled off to a museum. All three YF-12As had lives beyond their interceptor days. The first airframe was seriously damaged during a landing mishap in 1966. The forward fuselage was a loss, but the rear fuselage was mated with the forward fuselage of an SR-71 static test unit. With a raised second cockpit this became the sole SR-71C trainer. It was not well loved... when it flew it wanted to fly sideways; the yaw apparently due to the two airplane halves not being precisely aligned. It only flew training flights when the SR-71B trainer was down for maintenance. The SR-71C ended up at the museum at Hill Air Force Base in Utah.

The second and third YF-12As entered the service of NASA in 1969, performing a range of scientific missions from sonic boom testing to materials studies to high-altitude air sampling. The third YF-12A was lost in an in-flight fire; both pilots ejected but the airplane crashed. The second YF-12A served NASA until 1979 when it flew to Wright-Patterson Air Force Base for display at the then-named United States Air Force Museum.

While NASA had access to the YF-12A, proposals were made for modifications to the airframe and for interesting external payloads. One such payload was as a 1973 concept for a 'flying wind tunnel', carrying an inlet test article with the appearance of a third engine. The nacelle carried above the rear fuselage did not contain an actual engine, but focused on a 10-square-foot capture area inlet. The nacelle and inlet were similar to those of the Boeing 2707-300 supersonic transport that had been cancelled just a few years before.

Lockheed FB-12 with Four AGM-69 SRAMs

SCALE 1/100

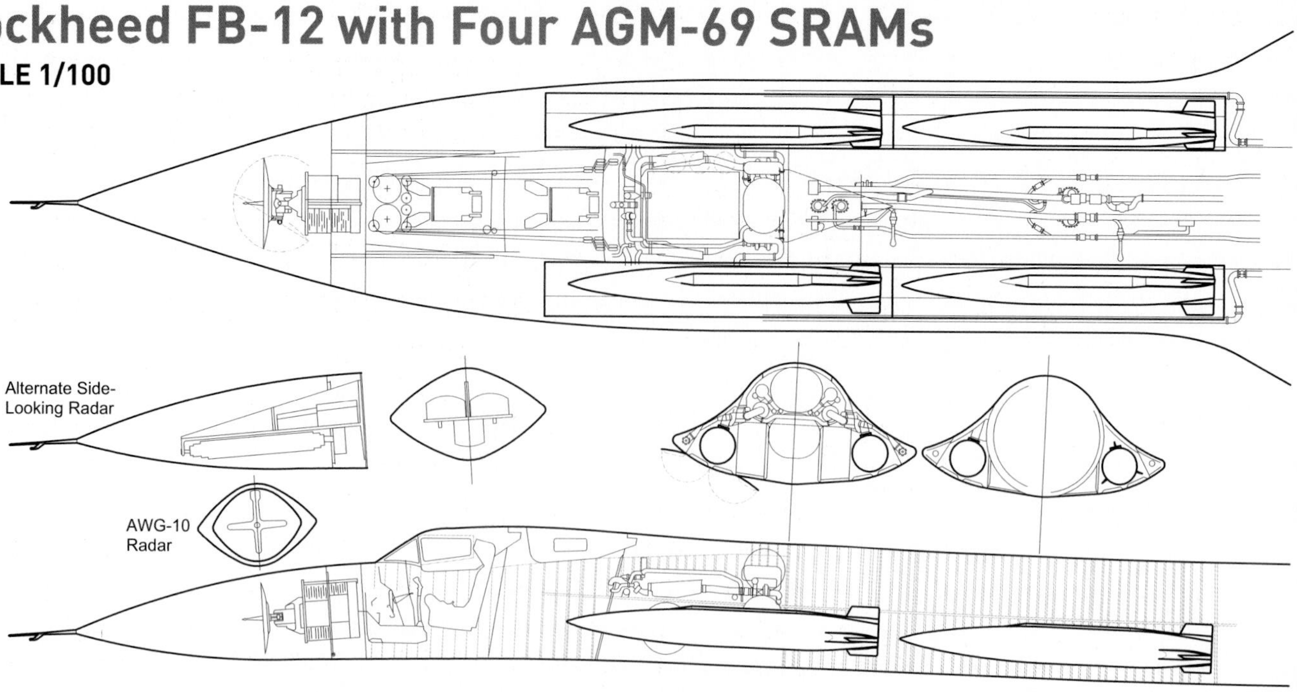

Alternate Side-Looking Radar

AWG-10 Radar

Lockheed FB-12, 2 SRAM, 2 AIM-7E/F

SCALE 1/65

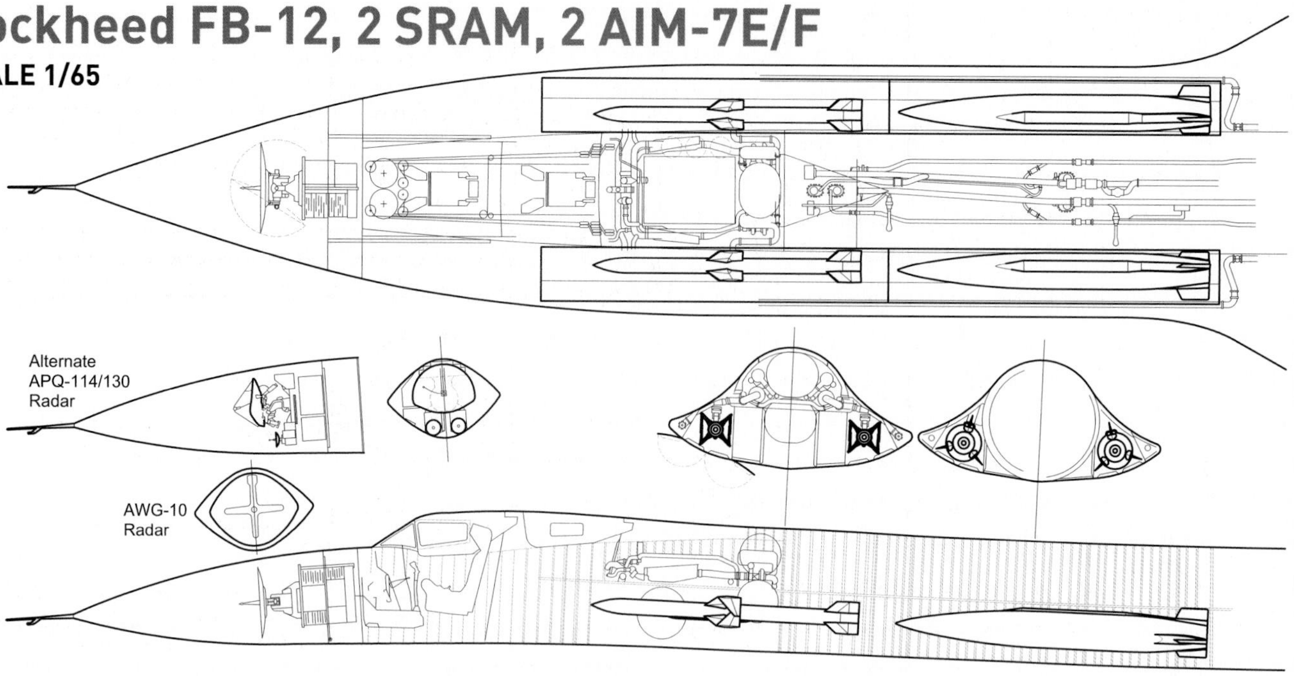

Alternate APQ-114/130 Radar

AWG-10 Radar

Lockheed YF-12A, 4 AIM-7E/F

SCALE 1/100

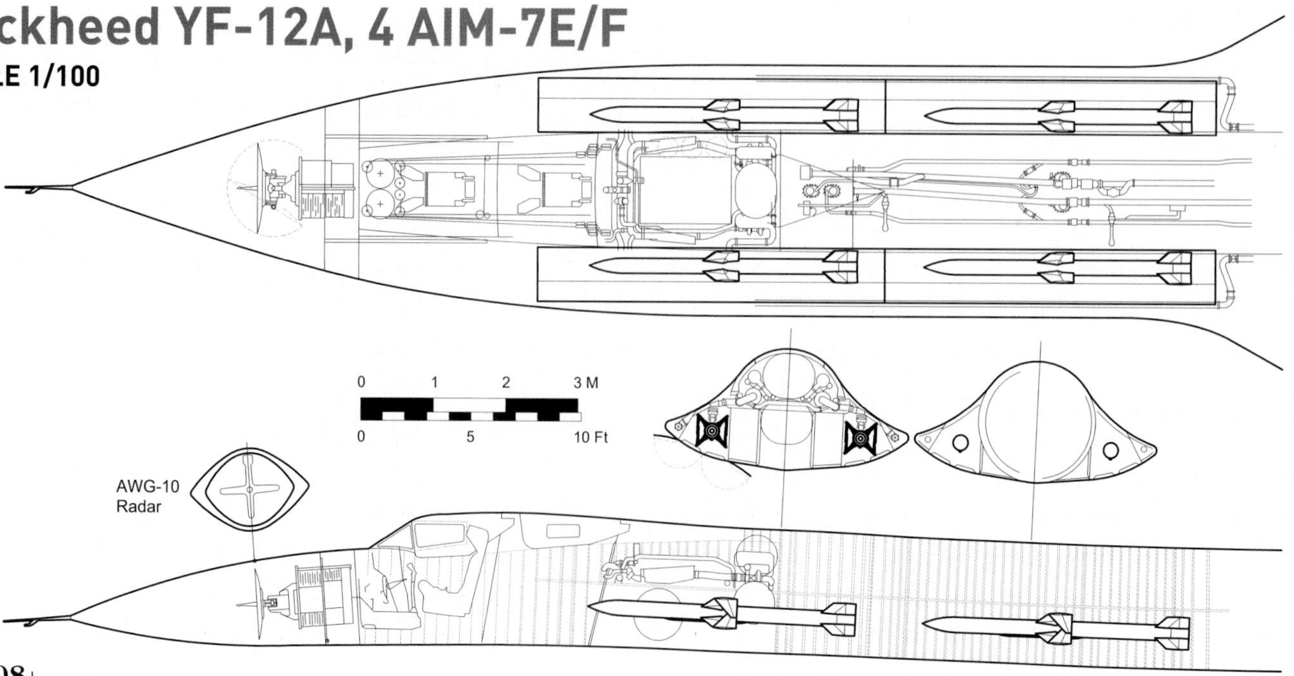

0 1 2 3 M

0 5 10 Ft

AWG-10 Radar

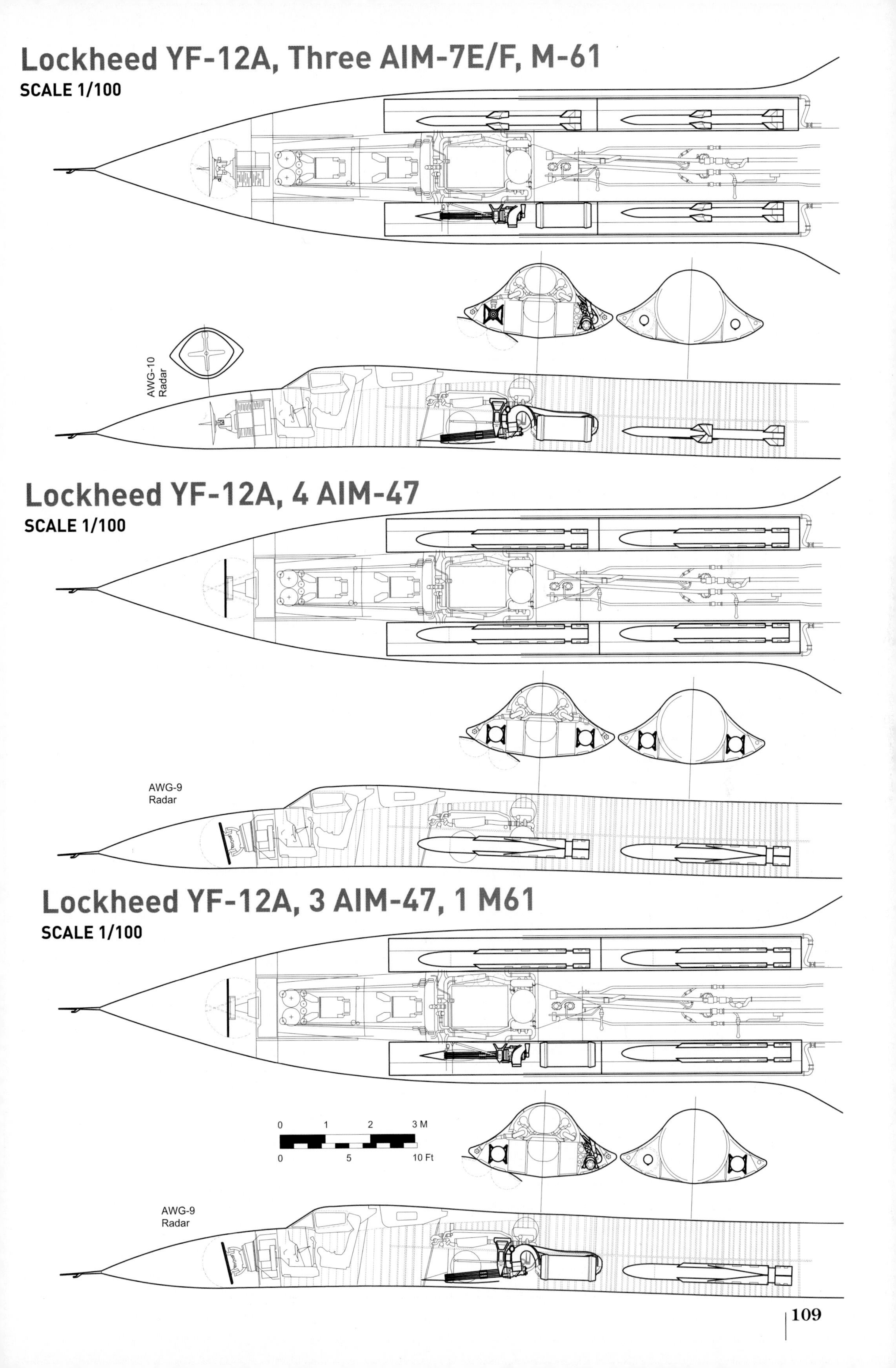
Lockheed YF-12A, Three AIM-7E/F, M-61
SCALE 1/100
AWG-10 Radar
Lockheed YF-12A, 4 AIM-47
SCALE 1/100
AWG-9 Radar
Lockheed YF-12A, 3 AIM-47, 1 M61
SCALE 1/100
0 1 2 3 M
0 5 10 Ft
AWG-9 Radar

6 CHAPTER The SR-71

The A-12 was designed specifically to be a world beating reconnaissance aircraft, but Kelly Johnson realized early on that with some modifications it could form the basis of a 'Universal' aircraft… recon platform, interceptor, bomber. Part of the drive for this was the fact that the A-12 was for the CIA, and its requirement for aircraft was strictly limited. The US Air Force, on the other hand, could potentially want hundreds of aircraft based on the 'Universal'. In March 1962, the Air Force gave a contract to Lockheed to study alternate roles for an A-12 'Universal'. By April, mockups of the forward fuselages of the 'R-12' Universal aircraft and the 'RS-12' strike variant were built.

The R-12 design differed from the A-12 in having a slight fuselage stretch (in particular an extension of the fuselage tailcone), allowing for additional fuel. The increased length affected longitudinal stability, so the chines near the nose were widened for a blunter planform. The Q-bay was replaced with a second pressurized permanent crew compartment, and the secondary bays added to the A-12s chines almost as an afterthought became major payload spaces along with the use of the nose to contain the primary recon equipment. The SR-71 had more space in the chines for payload, though not as voluminous as the missile bays on the F-12 as the interceptor had deepened the chines.

The aircraft aft of station 715, though, would not be greatly altered. But while the Air Force expressed interest, by December 1962 no official decision had been made to proceed. Over the next few months Lockheed and the Air Force went back and forth on the subject of the payload weight the aircraft would carry, ranging everywhere from 1500-4300lb. Finally in February 1963 the Air Force ordered six aircraft with a promise of ordering 25 more in the following summer.

Those first six aircraft were in fact part of the original A-12 order for the CIA, thus splitting the bill. By extending the production, down the line cost per aircraft was reduced; but the existence of a second variant considered to be potentially superior meant that the original run of A-12s would be the only run of A-12s. The first mockup review for the finalized R-12 took place in June 1963. It was considered successful and another successful mockup review occurred in December 1963. At the time the RS-12 was still planned, though undefined. It was to be armed with a Hughes missile (possibly a nuclear ground-attack version of the GAM-9/AIM-47). Installing this would have required structural changes. By March of 1964, construction of the first R-12 aircraft was under way.

In late July 1964, an important change came to the R-12 programme. President Johnson gave another press conference, this time announcing the existence of the 'SR-71' programme, accurately describing a high-speed reconnaissance aircraft. The 'SR-71' designation seemed to come out of nowhere and has for decades been the subject of dispute and supposition. In the same speech President Johnson made reference to the 'RS-70', the then-planned reconnaissance version of the B-70 Valkyrie. Many have assumed that Johnson simply transposed the R and the S in his head, but the actual prepared text of his speech shows that it was typed down as 'SR-71'. The '71' obviously followed after the B-70 and the 'SR' was, reportedly, General Curtis LeMay's preference for a 'Strategic Reconnaissance' designation, rather than 'Reconnaissance Strike'.

The first of the now-designated-SR-71s was delivered to Palmdale in late October 1964, beginning engine test in December. On December 22, it flew for the first time. As the SR-71 was, thanks to President Johnson, public knowledge, there was no need to conduct all of the test flights from Area 51. As the subsequent testing and operational history of the SR-71 has been told often and adequately in a number of other sources, there's no need to go over it again here.

Along with the addition of another crewmember, the main advantage of the SR-71 over the A-12 was the fact that it was a multi-sensor platform. The A-12 carried a single camera, while the SR-71 could carry multiple cameras as well as side looking radar and electronic surveillance equipment. The A-12 had an advantage in being able to carry a larger and more capable camera, but the SR-71's abilities proved much more practical.

The SR-71 was capable of carrying a mix of an Optical Bar Camera (OBC), side looking radar, two high resolution narrow field optical Technical Objective Cameras (TEOC), a high resolution side looking radar mapping system and Electronic Intelligence Improved Program/Electromagnetic Reconnaissance.

The primary sensor location was the nose, ahead of the cockpit. The nose of the SR-71 was a swappable module, in which either the OBC or the side looking radar system would be permanently mounted. The OBC from ITEK was a high resolution panoramic camera that scanned from left to right; at operational altitude, the terrain covered was two miles long (in the direction of travel) by 36 miles to each side. Resolution using the 30in focal length lens was reportedly about 8in. With a 10,500ft reel of film, the camera could cover 2952 nautical miles of travel with 10% overlap between images, or 1476 nautical miles with 55% overlap. The camera operation was automatic, but the Reconnaissance System Officer could operate it manually.

At least three different side looking radar systems were used during the SR-71's operational lifetime, reflecting the ever improving nature of electronic technology. All of the systems were used for terrain mapping – collecting data on the geography of target regions. This was just the sort of thing that would be needed for terrain following radar systems used on bombers and cruise missiles. The first system was the Goodyear 'PIP', followed up by the Loral 'Capability, Reconnaissance' (CAPRE) system. CAPRE was capable of a resolution of 5ft from cruising altitude. By the 1980s, radar and computer systems had improved to the point that a resolution of 1ft was available with the Loral Advanced Synthetic Aperture Radar System (ASARS) system.

The noses were geometrically similar, but could be distinguished: the underside of the radar system nose was relatively featureless, while the OBC nose had two windows, one on either side of the underside centreline.

An additional change was made to the nose. As originally designed and built, the chines continued smoothly and unbroken all along the planform from the pitot tube at the nose all the way back to the wing root. But 'dents' appeared in the chines about 20 centimetres aft of the pitot tube. These were blisters containing receivers for the 'DEF A' threat detection system. They picked up radar transmissions from missiles or tracking stations; on the underside of the nose, slightly aft and inboard of the blisters, transmitters sent out jamming signals.

The technical objective cameras were made by Hycon and were carried in pairs, one each in the left and right aft mission equipment bays. These were functionally somewhat similar to the camera in the A-12 in that they were traditional cameras (with a 48in focal length) using long spools of film. In the case of the TEOC, they could be pointed anywhere from straight down to 45° to the side. When looking straight down, they imaged a square region 2.4 nautical miles on a side, with a resolution of 4-5in. When looking off-axis, the region viewed expanded to a five by six nautical mile area, 14 nautical miles to the side of the aircraft. Each TEOC carried enough film for 1428 nautical miles of coverage. The cameras were automatically operated, their programming controlled by the aircraft's navigation system. Alternatively the Reconnaissance Systems Officer could control them manually.

The Fairchild Terrain Objective Camera (TROC) was a smaller, lower resolution camera located just forward of the nose gear bay. This camera did not serve a reconnaissance role so much as a political one; it showed the path the aircraft took, and could be used to prove that it did not overfly such-and-such a region if a diplomatic complaint was made. Of course, if the SR-71 *did* overfly that region, the TROC footage would likely not be shown.

In order for the SR-71 to know where it was, it used an astro-inertial navigation system. In the years before the advent of the Global Positioning System, this was the best system available for long-range military aircraft. A quartz window on the top of the fuselage, located aft of the RSO cockpit and fore of the refuelling receptacle, permitted an optical sensor to look up and out, scanning for at least two of a catalogue of 61 stars. By knowing where in the sky those stars were, and having a very accurate chronometer good to 1/100 of a second, it was possible for the automated system to know where on earth it was to within around 300ft. This worked well during straight and level flight; during banking, though, the system would be disrupted. The astro-inertial navigation system did not need input from the pilot or reconnaissance systems operator for normal functioning, though such input would be useful for times when stars were obscured (by clouds or a KC-135 refuelling boom, for instance). In the event that the astro navigation system failed, a gyro-inertial system was included. Position errors would of course mount up during the course of the mission.

For an aircraft operating about 80,000ft, stars would be visible at all hours of the day. At that altitude the sky is black, not blue; the aircraft appears to be in space.

SR-71B Trainer

In January 1966, the seventh SR-71 completed and delivered was the first SR-71B trainer, with a gross weight of 139,200lb carrying 59,000lb of fuel.

Configured like the A-12T with a raised second cockpit for the instructor, this aircraft was conveniently the first SR-71 to enter operational US Air Force inventory. Two SR-71B trainers, tail numbers 64-17956 and 64-17957, were delivered. Along with the raised second cockpit, they featured the outboard ventral fins of the YF-12A to offset the yaw moment produced by the added area of the raised cockpit. The raised cockpit, while the same idea as that of the A-12T, was configured quite differently, being substantially longer and sleeker.

The instructor's cockpit was, unlike the station of the reconnaissance systems operator in a standard

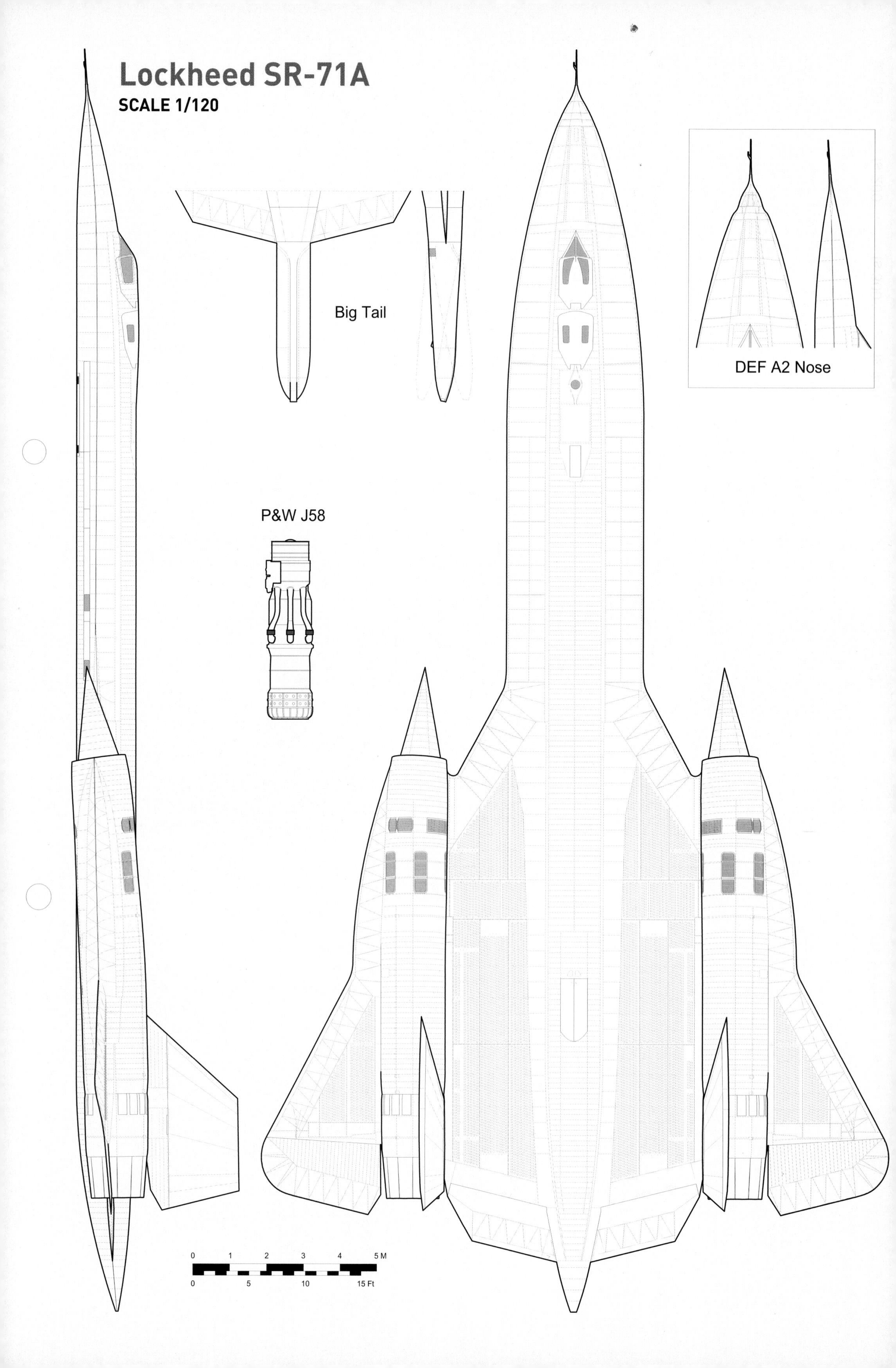

Lockheed SR-71A
SCALE 1/120
Big Tail
DEF A2 Nose
P&W J58
0 1 2 3 4 5 M
0 5 10 15 Ft

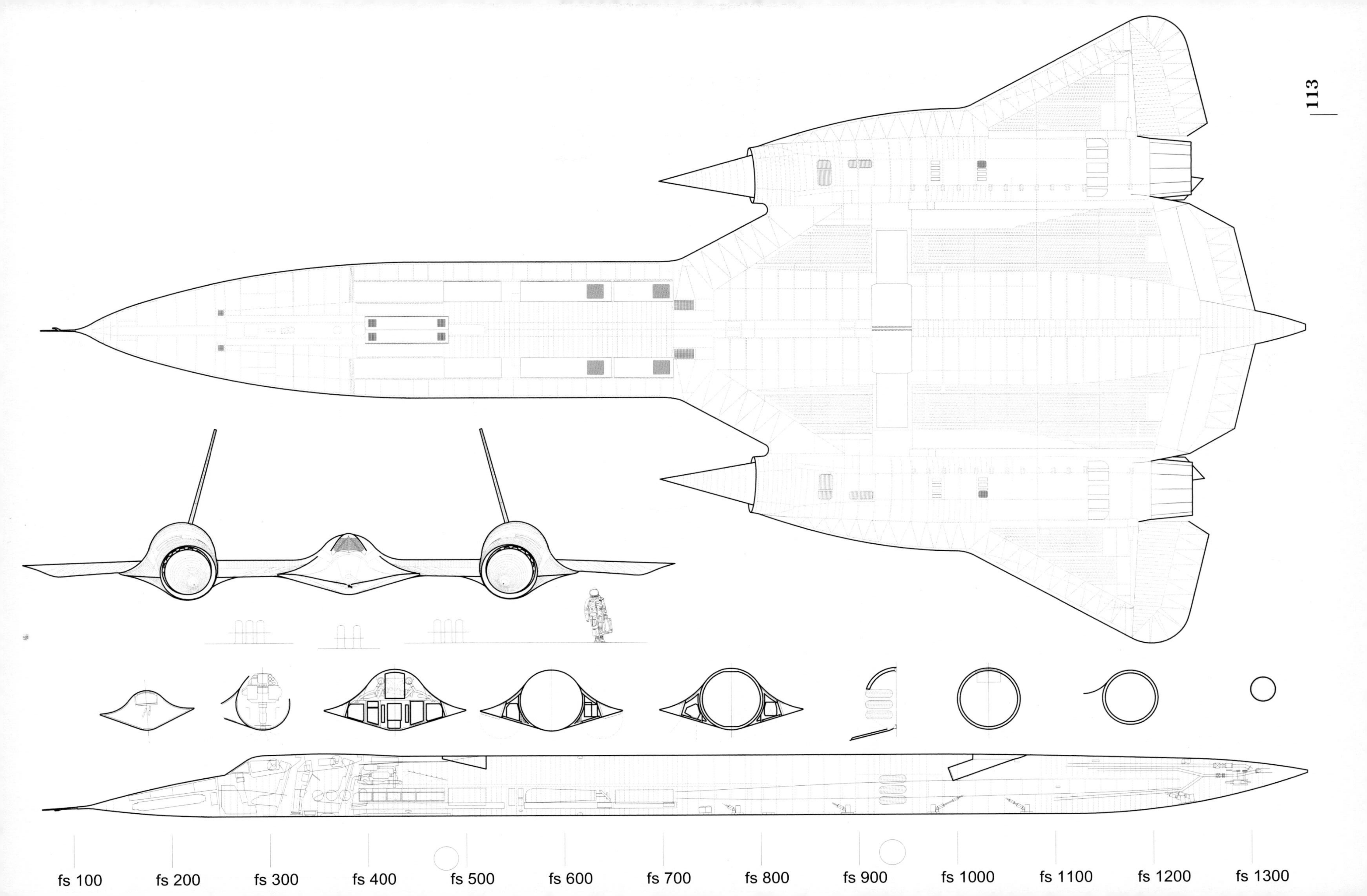
fs 100
fs 200
fs 300
fs 400
fs 500
fs 600
fs 700
fs 800
fs 900
fs 1000
fs 1100
fs 1200
fs 1300

SR-71, fitted out to allow for control of the aircraft in the event the trainee pilot proved incapable. However, it did not have a full set of controls; the electrical system, for instance, was solely controlled from the forward cockpit. Flying the aircraft from the rear cockpit was apparently something of a chore. Fortunately, trainee pilots were all experienced experts long before they took to the air in the SR-71B. Not only were they among the best pilots the Air Force had to offer in the first place, they underwent a substantial training process on the ground using advanced (for the time) simulators.

SR-71B 64-17957 suffered a double-generator failure in January 1968 while over Montana. All possible landing sites between there and Edwards Air Force Base were ruled out due to winter weather conditions and unfortunately, seven miles short of the runway, the aircraft crashed. Fortunately both pilots safely ejected. The other SR-71B went on to fly for NASA and is now on display in Kalamazoo, Michigan.

Big Tail

Numerous studies were carried out through the early years of the SR-71's life to figure out how to carry the side looking radar and the optical bar camera at the same time, rather than having to choose by swapping out noses. Most studies involved attaching external pods to the underside of the rear fuselage containing either camera or radar systems. These pods though, would add weight and drag; none were implemented. One system that was built and flown, though, was the 'Big Tail' modification.

Unlike many of the code names attached to the SR-71, Big Tail was surprisingly accurate and descriptive. First flown in December 1975 on SR-71 61-7959, the Big Tail was exactly what it said it was… a nearly 9ft long extension to the aft fuselage. It added 1273lb, but also added 864lb of payload and 49 cubic feet of payload volume with almost no additional drag. The tail extension was long enough to cause a problem as the plane rotated on takeoff though, so the Big Tail was hinged to angle upwards 8.5° to prevent runway strikes. After takeoff it would angle straight back; after landing it would angle downwards -8.5° to provide clearance for the drag chute.

The Big Tail was initially meant to carry an enhanced electronic countermeasures suite. At the time there was concern that advances in Soviet surface-to-air missiles could mean that the SR-71 was vulnerable to an attack from the rear, but this threat never truly materialized. While the rearward ECM system was installed in Big Tail, it was primarily used to demonstrate a tail-mounted optical bar camera, freeing the nose compartment for the side looking radar. It was also used to test a satellite uplink system to send near-real-time data to local commanders and the Pentagon, a technology that is now pretty well standardized.

The last flight of 959 and the Big Tail was in October 1976. It was never used operationally.

B-71

The existence of the SR-71 bomber designs is well attested, if unfortunately currently rather poorly documented. As the SR-71 began life as the R-12 derivative of the A-12, the RS-12 strike version was designed right along with it. Sadly, while it appears that considerable effort went into the RS-12 design – later dubbed the B-71 – little is really known.

The best known design for a B-71 comes from fragmentary and poor-quality diagrams dated 1965. This B-71 carried six gravity bombs geometrically somewhat similar to the bombs carried by the 1961-vintage B-12. But instead of putting the bombs into a single bomb bay, the B-71 put each bomb into its own bay in the chines, three in a row on each side. The purpose of the B-71 would not be to lead the charge into war as the North American B-70 was intended to. Instead, it would 'mop up' Soviet missile sites after the first wave of a nuclear war had been fought. The B-71 would fly into soviet territory and look for surviving missile sites, using the side looking radar. If a nuclear system was discovered, the B-71 would drop a thermonuclear gravity bomb onto it.

Curiously, it appears that the B-71 would actually have had a shortened forward fuselage. This is odd, especially since this meant that the forwardmost bomb bays required fairings be added to the top of the chines, as the chines were thinner closer to the nose.

YF-12C Studies

After the loss of the third YF-12A while loaned to NASA in 1971, an SR-71 was provided to fulfill a similar role. For reasons that doubtless made sense to the kind of people that such ideas make sense to, this SR-71A was dubbed the 'YF-12C' while in NASA service. The claim is that this somehow would have prevented the public from knowing that NASA was operating the USAF spyplane, rather than the ex-interceptor.

While NASA operated the 'YF-12C', Lockheed made a proposal for modifying the aircraft to carry a range of external payloads. Many of these involved the addition of an adjustable boom to the back of the rear fuselage. The 'back-hoe' as it was called would change slightly from design to design.

One concept called for the use of the back-hoe or a fixed pylon to demonstrate supersonic weapons carriage and release. The fixed pylon idea would be used for captive carry only, since the missiles would be located above the rear fuselage. The vertical pylon would carry one missile – such as an AIM-7 Sparrow or an AIM-9 Sidewinder – on each side, the missiles mounted to smaller horizontal pylons. The secondary

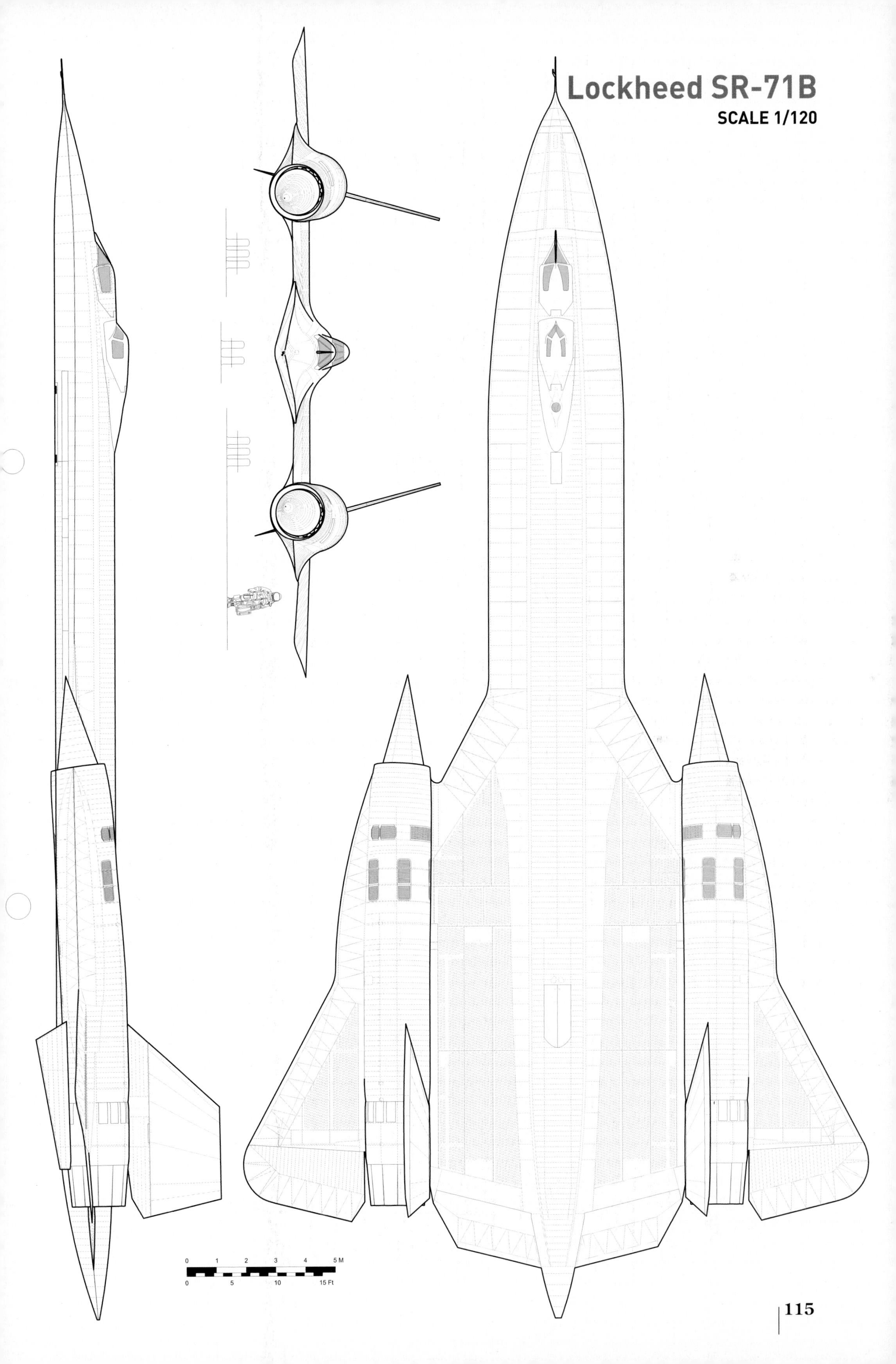
Lockheed SR-71B
SCALE 1/120
0 1 2 3 4 5 M
0 5 10 15 Ft

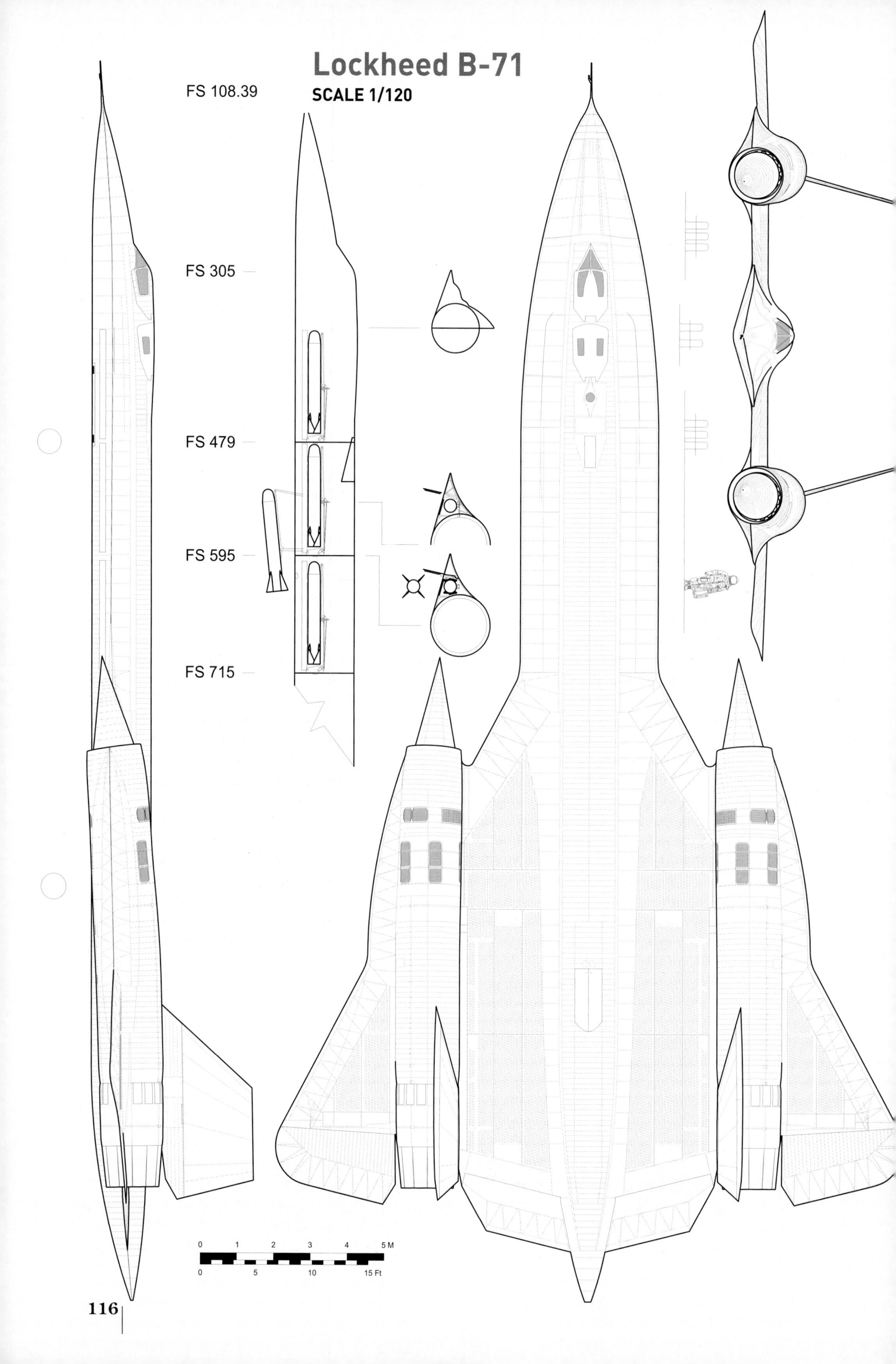
Lockheed B-71
SCALE 1/120
FS 108.39
FS 305
FS 479
FS 595
FS 715
0 1 2 3 4 5 M
0 5 10 15 Ft

pylons would be able to pitch up and down depending on current airflow conditions; the system was meant to test not just the missiles, but the effect of having asymmetrical weapons loads.

The back-hoe would be used to demonstrate high-Mach weapons separation. In this case, the pylon would not be fixed, but would instead have a variable boom. For normal flight the boom would hold the missile to the rear in a minimum drag configuration, but for launch the boom would angle down and release the missile in a nose-down attitude. It was expected that at least initially the missiles would be inert and simply dropped; live launches would wait until the post-drop flight characteristics of the missiles were well understood.

It was thought that the back-hoe concept could also be designed to demonstrate some of the technologies useful for supersonic in-flight refuelling. This somewhat terrifying concept dates back to the 1950s; if a long-range supersonic-cruise aircraft didn't need to slow down to rendezvous with a tanker aircraft, in theory a great deal of both time and fuel could be saved. While the theory sounds good, the realities of close formation flying at supersonic speeds – not to mention how refuelling systems such as drogue-equipped hoses deal with supersonic airflows – remain murky.

This concept would have tested out at least parts of that thorny problem by equipping the back-hoe with a hose, drogue and reel. It would not have actually transferred fuel, there being no fuel pumps or connection to fuel tanks, but at least the response of the hose and drogue would have been observed. It's possible that actual rendezvous and even docking might have been attempted if testing had gone well enough.

At the time, one of the biggest aerospace development programmes in the United States was the Space Shuttle. Lockheed suggested that the YF-12C could provide useful aerodynamic data. The planform of the YF-12C and the Shuttle were vaguely similar, with similar sizes; thus there was some thought that flight characteristics of the Lockheed vehicle might be predictive of Shuttle characteristics. But more importantly, the YF-12C could be used to carry Shuttle models to high speed and altitude. These models could be permanently attached the aircraft, as a sort of flying wind tunnel; or they could be dropped.

One notion called for the use of the back-hoe to carry a one-tenth scale Shuttle orbiter behind the aircraft. The back-hoe would be a fixed, unmoving structure that would drop the Shuttle model behind the YX-12C. The model would be equipped with a Rocketdyne AR2 rocket engine and enough propellant to reach Mach. The AR2 is perhaps best known as the rocket engine that was added to the Lockeed NF-104A Aerospace Trainer, and burned kerosene fuel with 90% hydrogen peroxide oxidizer. The AR2 could produce 6000lb of thrust, which would do well to accelerate the model, which had a dry weight of 1730lb and a fuel weight of 540lb.

The adjustable back-hoe could alternatively be used to carry a complete 1/15 scale Space Transportation System model... Orbiter, External Tank and Solid Rocket Boosters. This model might not have had functional booster rockets, but to provide clearance for separation the end of the tail of the YF-12C would be chopped off. The adjustable back-hoe would not only lower the model for release, it would also angle it distinctly nose-down.

A similarly configured back-hoe would carry a 1/10 scale orbiter with an External Tank. In this case, both the External Tank and the Orbiter would be equipped with solid rocket boosters. The Orbiter would be fitted with the motor from an AGM-69 Short Range Attack Missile, while the External Tank would have a modified Lockheed A-92 'Avanti' solid propellant rocket motor. The Avanti would be fitted with a new vectorable nozzle. If launched at Mach 3.2, this combination of booster motors would be able to accelerate the Orbiter model to Mach 9. However, the total mass added to the YF-12C would prevent the aircraft from reaching Mach 3.2, so further study was called for. As with the full STS stack, the end of the YF-12C's fuselage would need to be cut short to provide sufficient clearance for separation.

Another one-tenth scale Orbiter model used a total of three SRAM motors... one in the body of the model, two side-by-side behind the model. This two-stage vehicle would also be able to reach Mach 9; but unlike the Avanti-boosted version, this one appeared to have a low enough weight for the YF-12C to reach the required Mach 3.2 launch velocity.

As well as carrying models with the back-hoe, the YF-12C could carry models elsewhere. Models could be carried atop a pylon mounted on the centreline of the rear fuselage, using a similar setup to the inlet test proposed for the NASA YF-12A. Aircraft and other aerospace vehicle models would be attached to 'stings' fitted to the top of the pylon, articulated to pitch the model up to 30° upwards to study aerodynamics at various angles of attack. Along with a 1/10 scale Shuttle Orbiter, the available diagrams show, somewhat strangely, a 1/12 scale YF-12C and a possibly generic 'future store', presumably a missile of some kind. The YF-12C could also carry the 1/10 scale Orbiter beneath the left wing just outboard of the fuselage using a forward-swept pylon. Clearance between the bottom of the Orbiter model and the runway would be limited. This would be used for drop tests of an unpowered model.

One of the more surprising concepts was to use the YF-12C to test the aerodynamics of the Rockwell B-1 bomber. As was proposed for the Shuttle Orbiter, a 1/10 scale B-1 model would be carried by an adjustable back-hoe. The variable-sweep wings of the B-1 would be represented by replaceable fixed-sweep wings, from 15° sweep which would be the bombers takeoff wing sweep, all the way back to 67°. Exactly what value there would be in Mach 3.2 test data for a model of an aircraft barely capable of Mach 2.2 is unclear; murkier still is the value of any supersonic data for the configuration with the wings fully extended as if for takeoff. In any event, the model could be dropped for a (presumably) unpowered glide; the cockpit escape capsule that was built into the B-1A – but eliminated on the B-1B – could be demonstrated by jettisoning the scale equivalent from the model while in flight.

HT-4 drone

Another Lockheed concept from 1971 was the HT-4 hypersonic drone, carried in much the same way as the D-21 drone. The HT-4 was a purely rocket-powered vehicle configured like a North American Rockwell hypersonic transport design. With a Pratt & Whitney RL-10 rocket, it could generate 15,000lb of thrust to a total weight of 14,800lb. Separation would occur at 77,000ft and above Mach 3. Unlike the D-21 though, the HT-4 would be assisted in separation with the use of a jack that would angle the drone 8° nose upwards, using aerodynamic lift to accelerate the drone upwards.

SR-71(Bx)

In 1976 Ben Rich of the Lockheed Skunk Works delivered a paper that briefly described the 'Bx', a straightforward modification of the existing SR-71 into a supercruising strike aircraft. This would be accomplished by putting AGM-69A SRAM missiles into the chines… not quite into the existing equipment bays, but rather into new longer bays in approximately the same positions. The available diagram of the SR-71(Bx) is distressingly low resolution and has a few internal contradictions, but it appears that the forward fuselage of the aircraft would be 'bent' upwards in the same manner as that of the FB-12, along with having the blunter chines and the rounded underside of the FB-12 to provide sufficient clearance for the missiles. The radar would be different, though it's difficult to make out the details; the nose would be recontoured and slightly lengthened to accommodate.

Further data on the SR-71(Bx) is lean; how rigorous the design effort for the SR-71(Bx) was is unclear. But

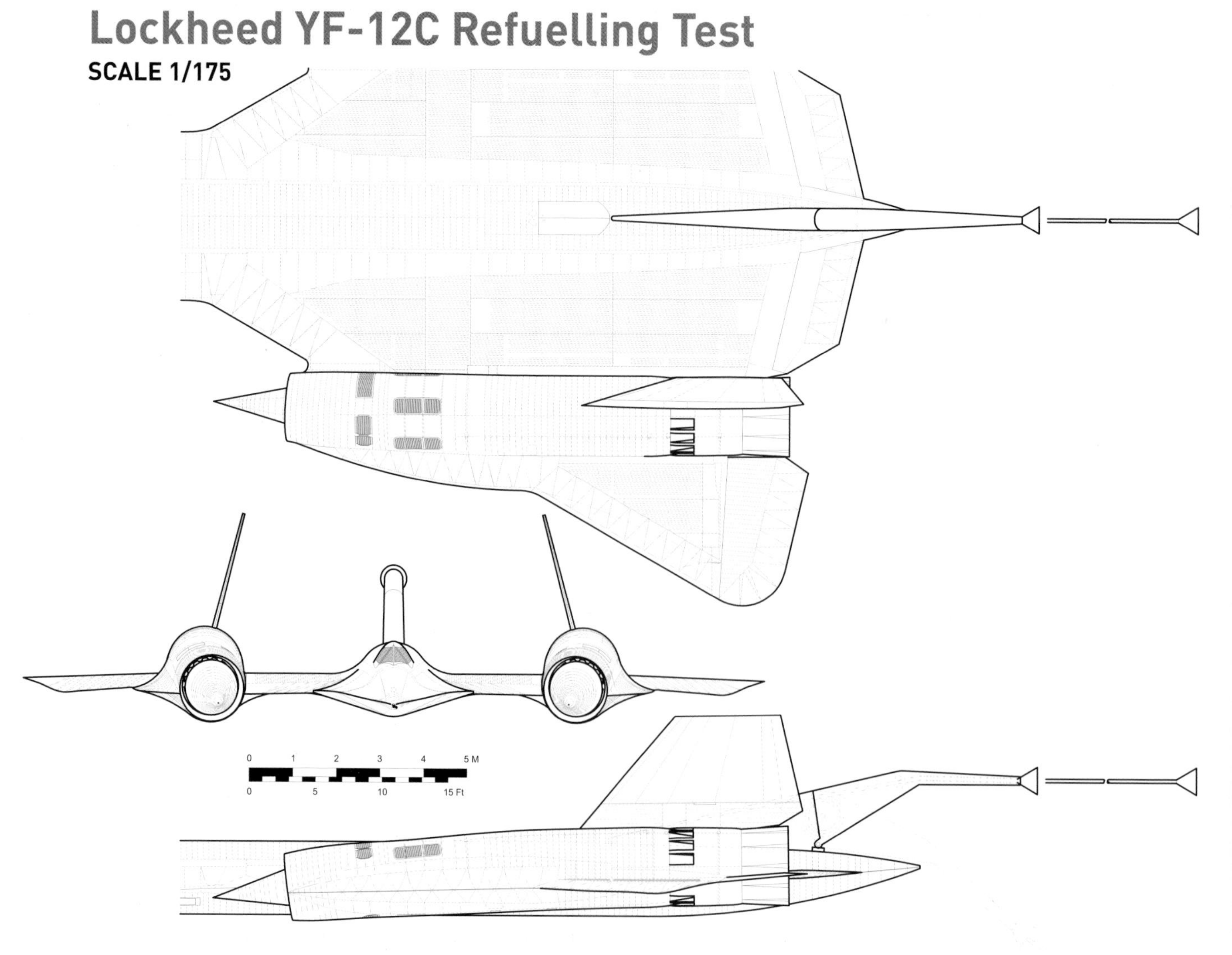

Lockheed YF-12C Flying Wind Tunnel

SCALE 1/175

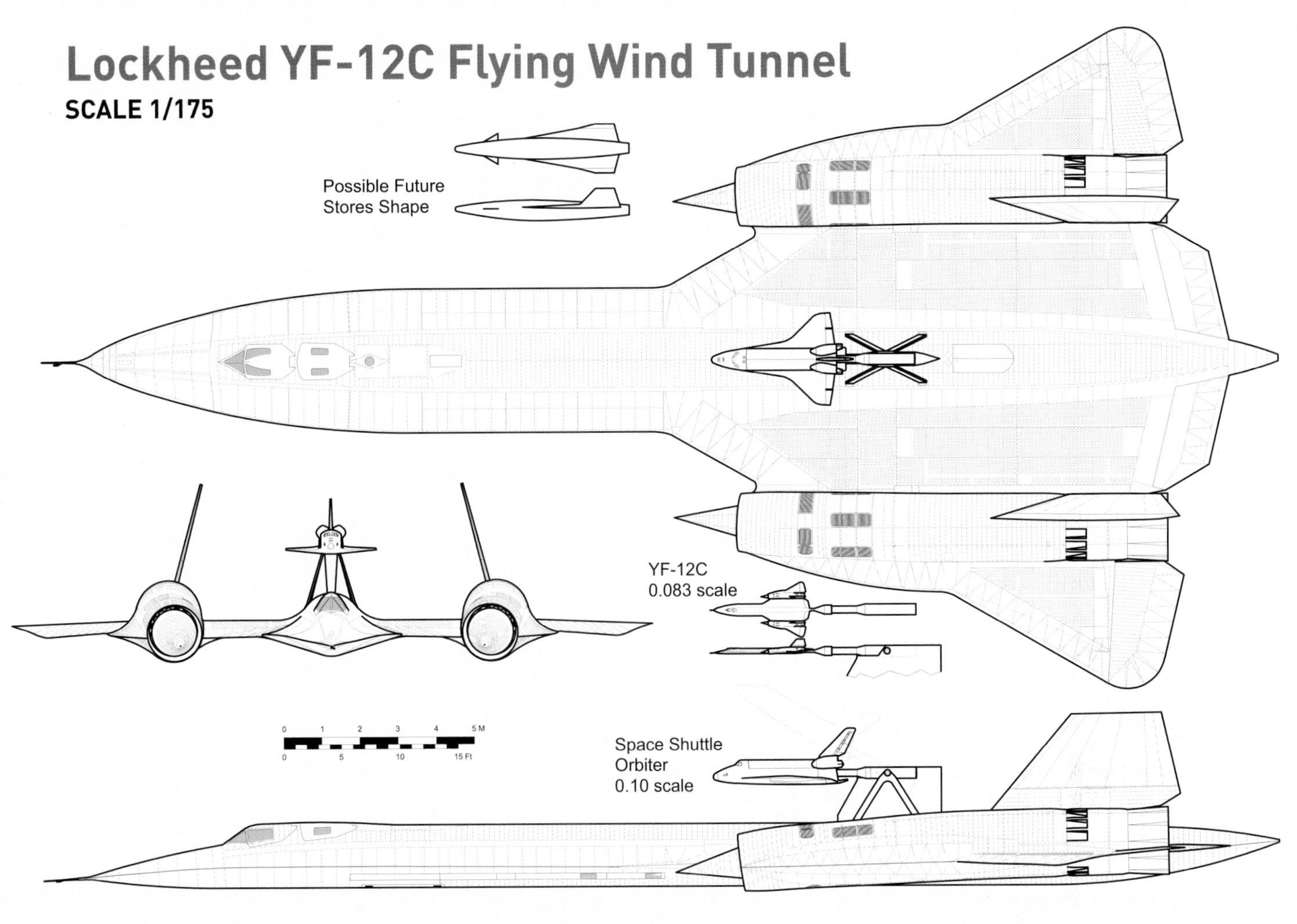

Lockheed YF-12C with Underwing Shuttle Model

SCALE 1/175

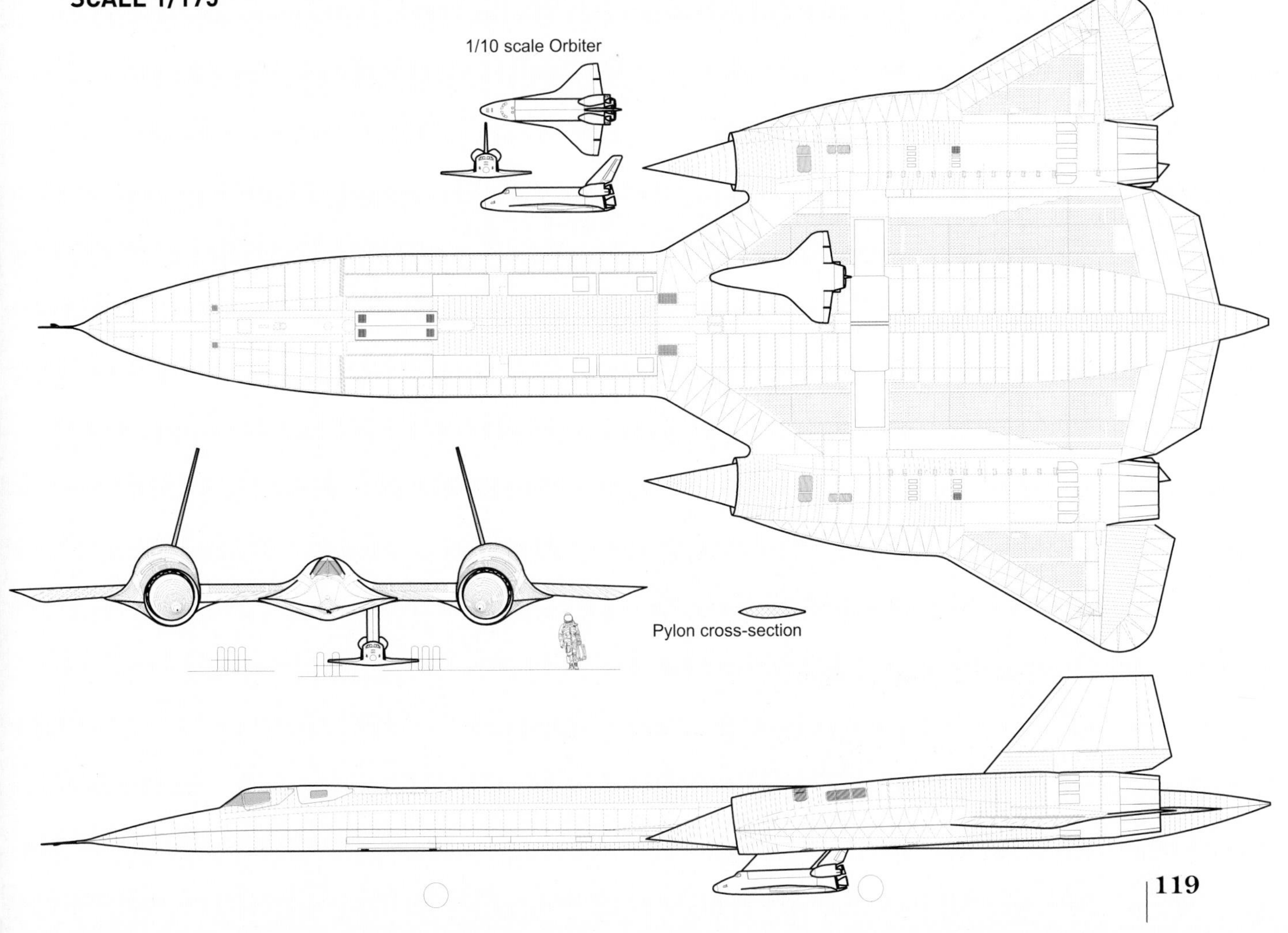

Lockheed YF-12C Shuttle Model with SRAM Boosters

SCALE 1/144

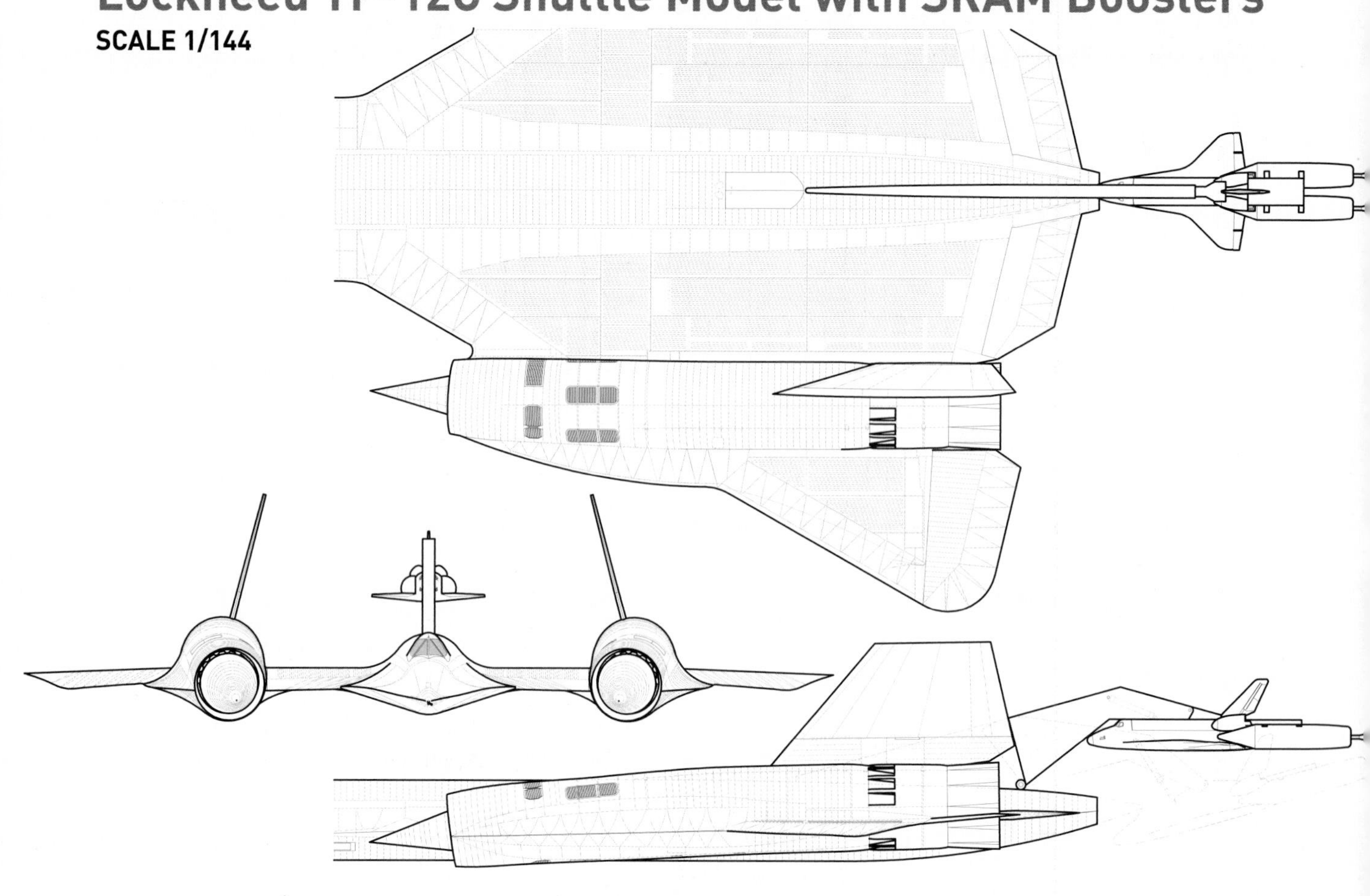

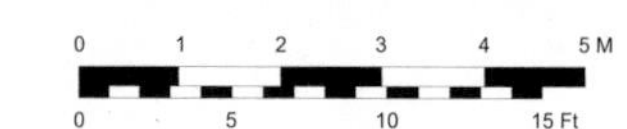

Lockheed YF-12C STS Stack

SCALE 1/144

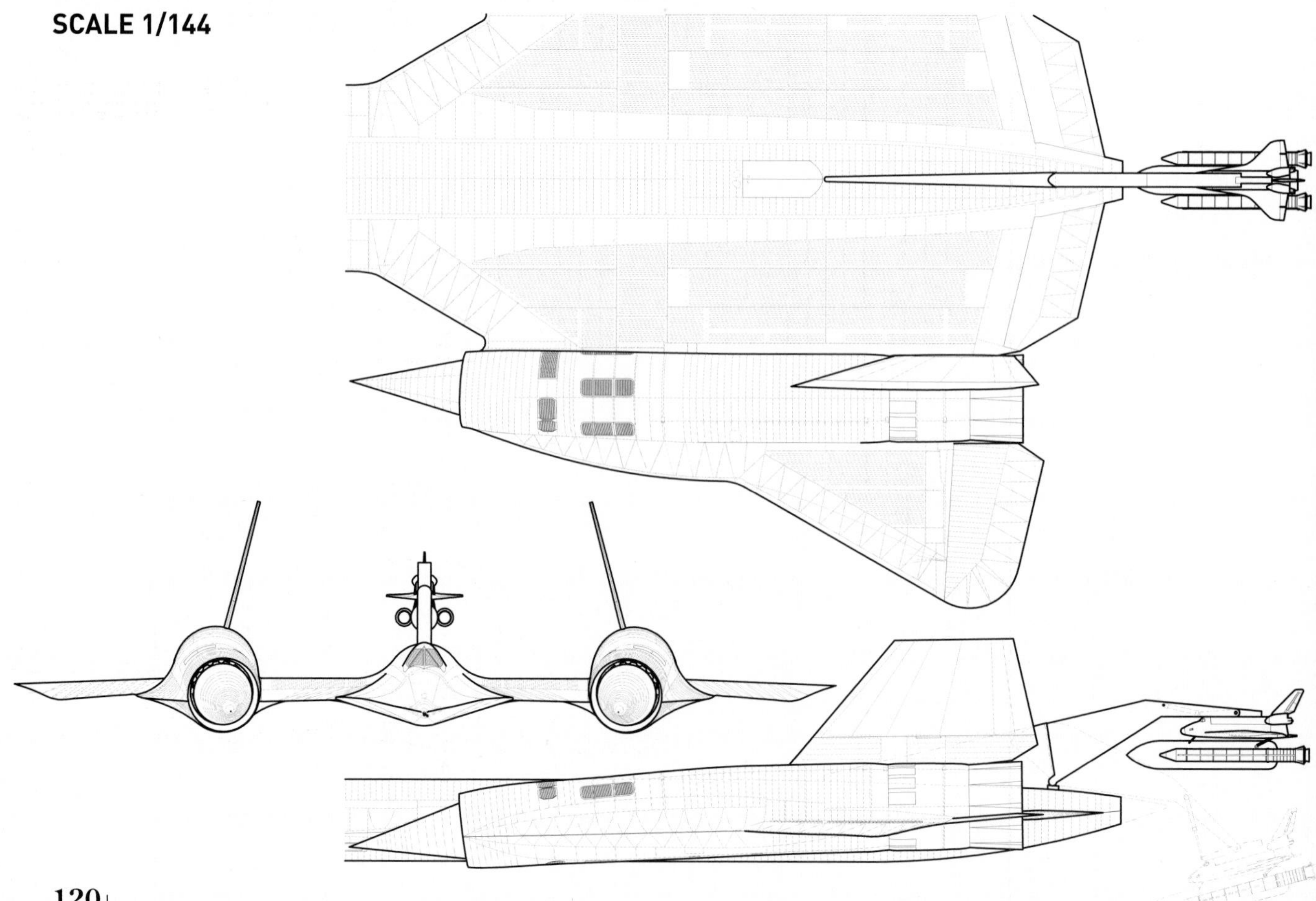

Lockheed YF-12C External Weapons

SCALE 1/120

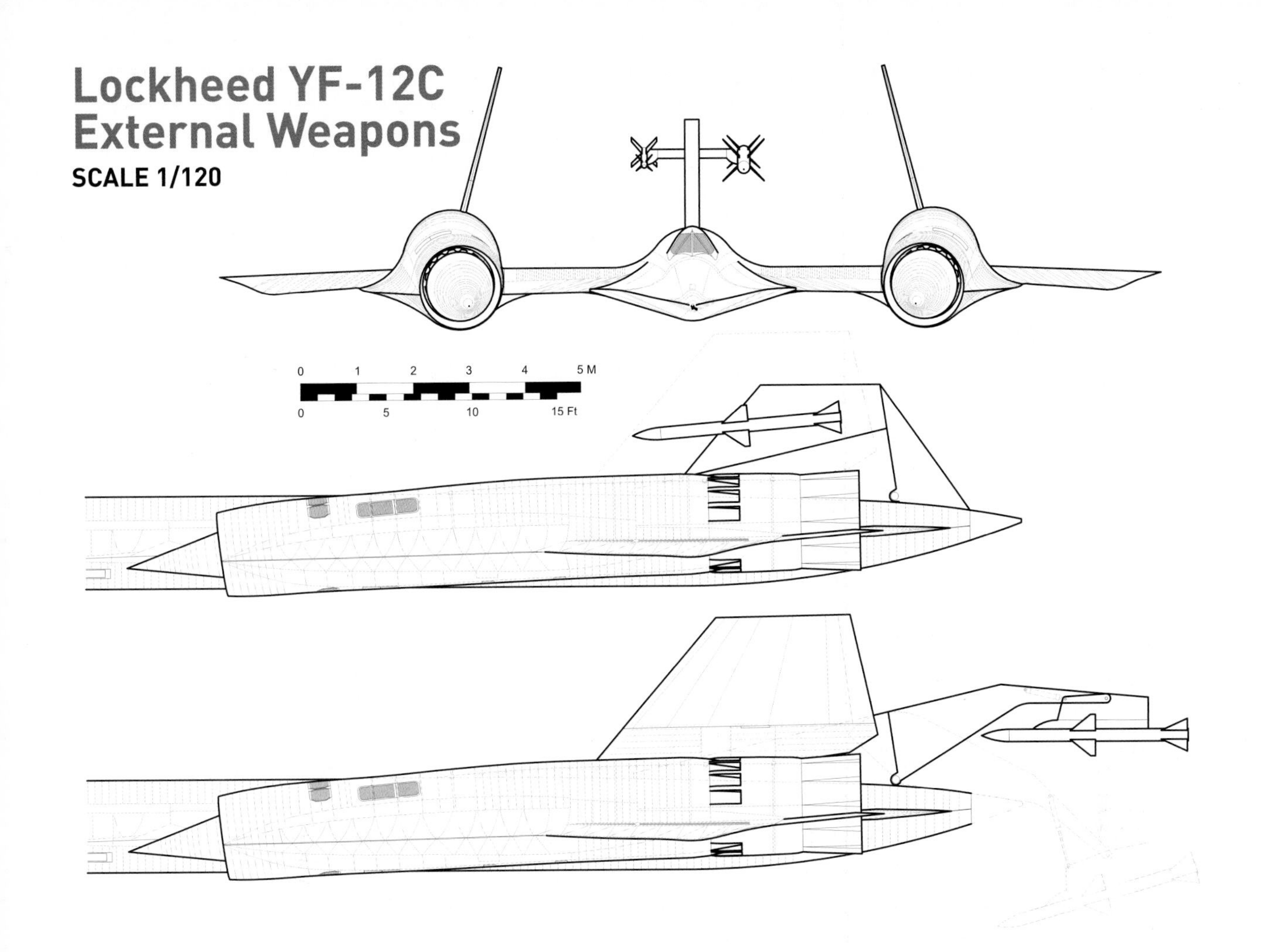

Lockheed YF-12C Fixed Aft Boom Orbiter Model

SCALE 1/144

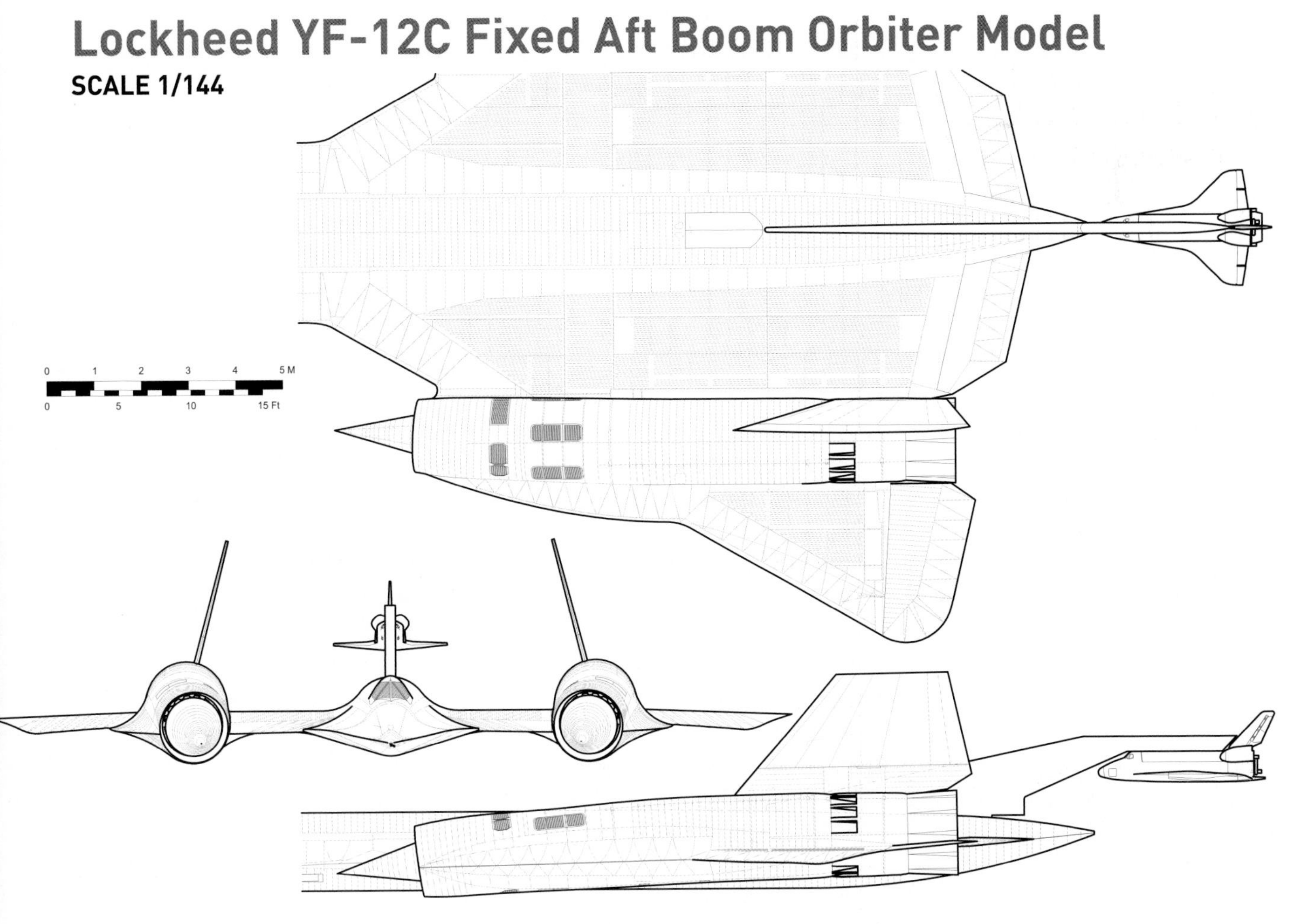

Lockheed YF-12C Shuttle Model with Avanti Booster

SCALE 1/144

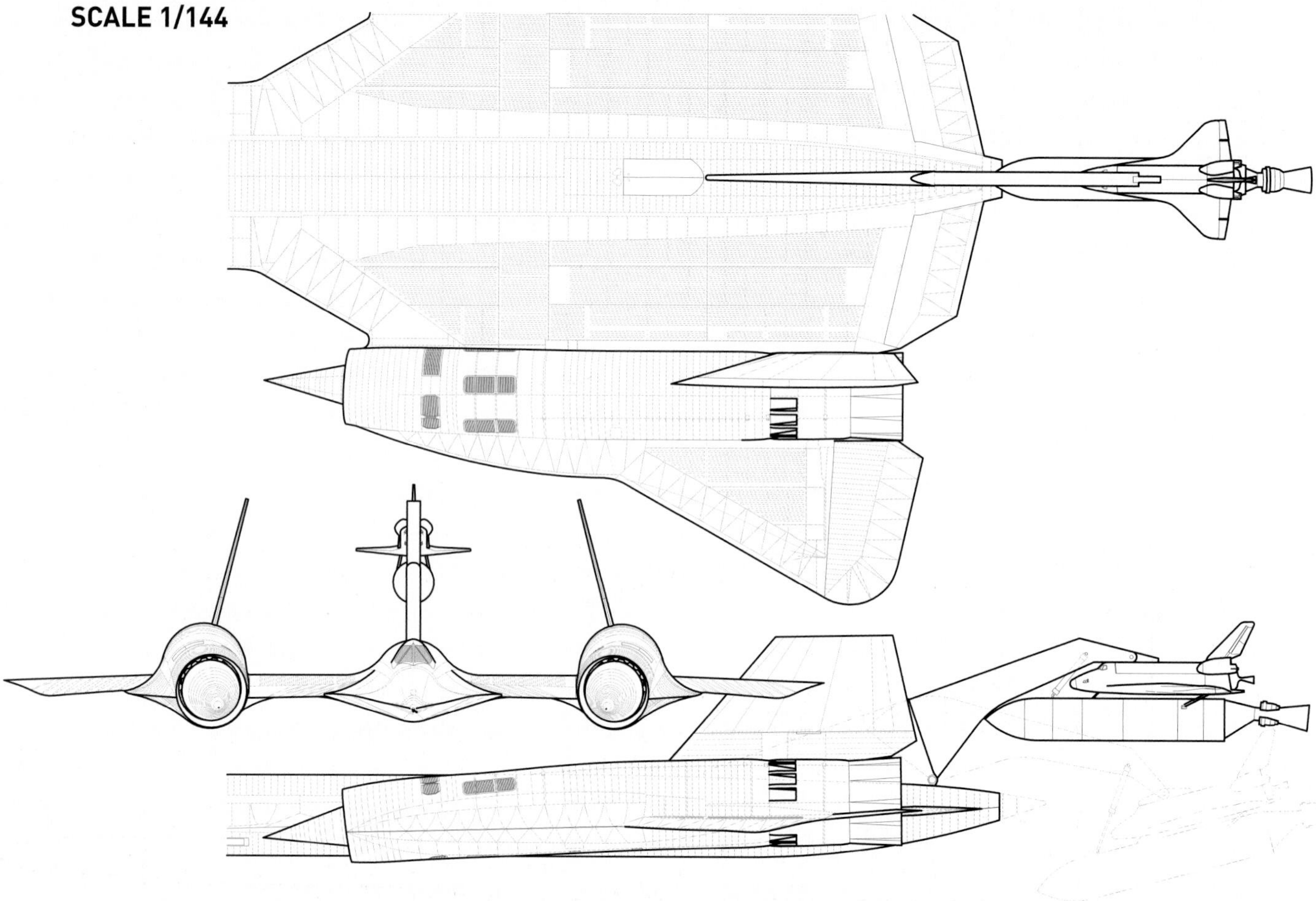

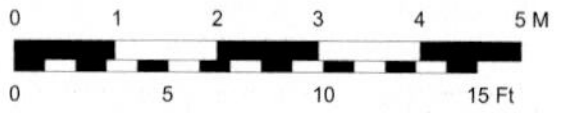

0 1 2 3 4 5 M

0 5 10 15 Ft

Lockheed YF-12C B-1 Model

SCALE 1/144

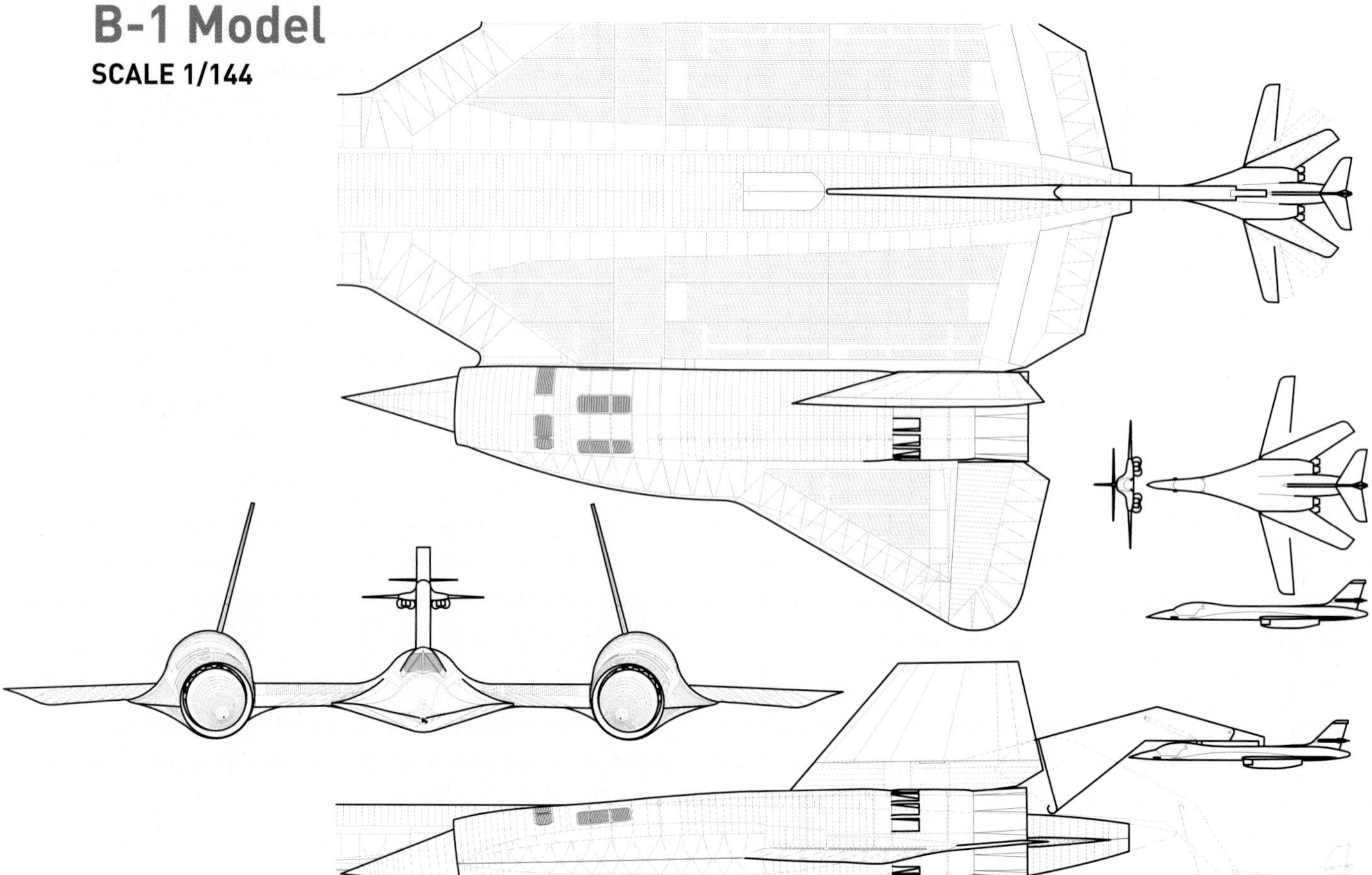

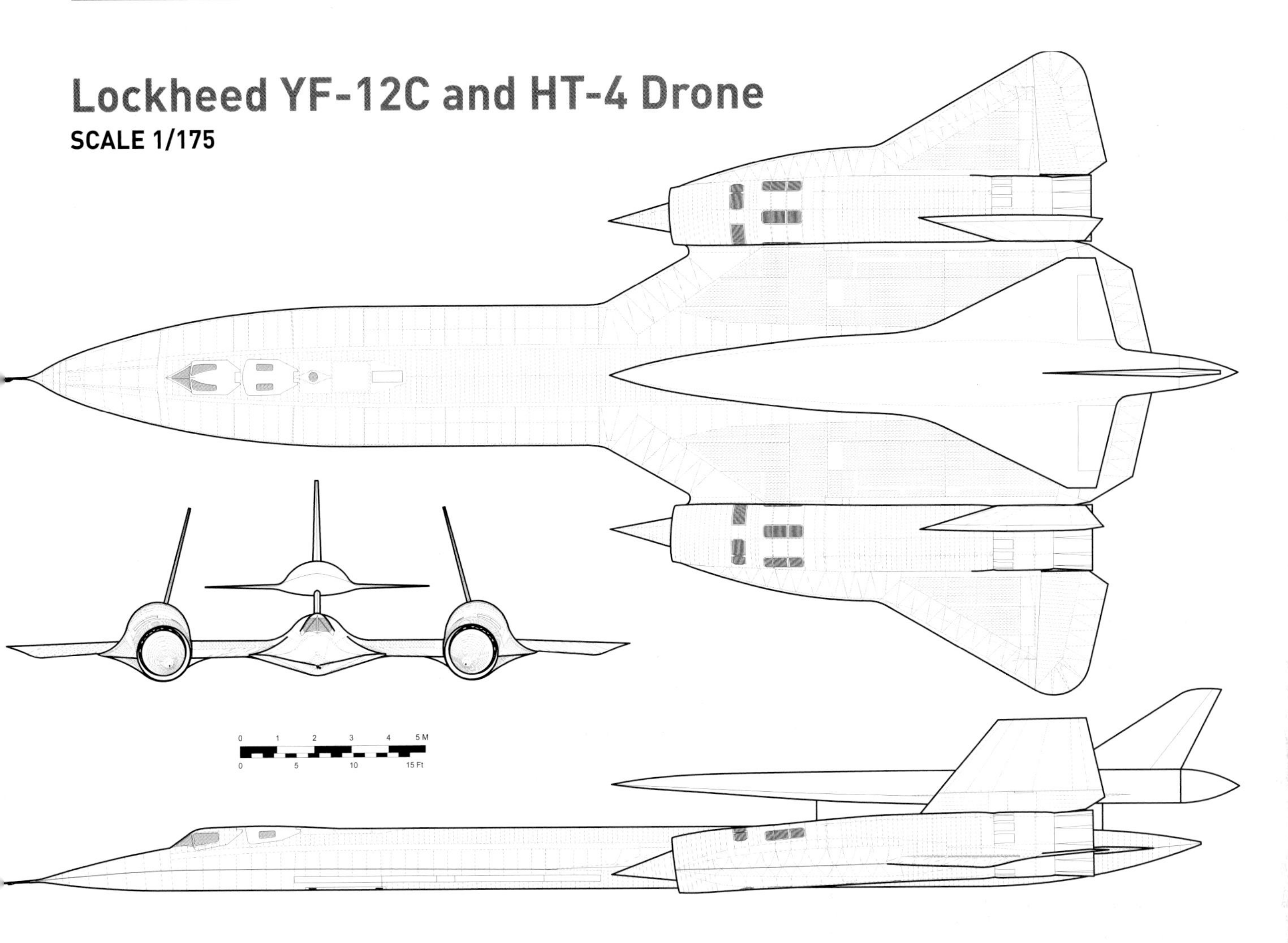

by launching at Mach 3.2 and at an altitude of 80,000ft, the range of the SRAM could be notably increased to around 500 nautical miles. As the SRAM was not a guided missile, accuracy was not spectacular… circular error probability at 300 nautical miles was 3600ft. Rich suggested that CEP could be cut in half with terminal guidance, but details on that were not given.

SR-71 I

In September of 1980, Analytic Services, Inc. (usually known as ANSER, a non-profit corporation formed in 1958 as a sort of 'think tank' to aid the US Air Force) produced a presentation on the use of the SR-71 as a long range interceptor. The SR-71 would be equipped with four AIM-54 Phoenix missiles – the inheritors of the AIM-47 Falcon legacy – and the AWG-9 radar for the purposes of reaching out and destroying Soviet AWACS aircraft. The presentation was unfortunately thin on details about configuration changes, though it noted that the nose would need to be re-contoured; options also included changing the engines to F100 or F101 turbofans and stretching the fuselage. The Phoenix missiles would be carried within the internal bays, but external carriage under the rear fuselage was also touched upon. The Phoenix missiles would require folding fins to fit within the SR-71 bays. This study seems to have been largely back of the envelope.

One of the latest known ideas for creating an interceptor variant of the Blackbird was the 'SR-71 I', proposed by Lockheed to the USAF in November 1982. This design called for the use of the new AIM-120 AMRAAM (Advanced Medium Range Air-to-Air Missile), an improvement over prior designs calling for the use of the hypothetical AIM-7E/F Sparrow. The AIM-120 was essentially the same size as the AIM-7E/F without requiring the complexity of folding fins. The Sparrow missile required that the target be illuminated by the radar unit in the aircraft, while the AIM-120 had its own radar transmitter. Thus it was a 'fire and forget' missile. A bonus was that since the aircraft's radar did not need to constantly track a single target until the missile hit it, multiple enemy aircraft could be tracked and targeted simultaneously.

In order to support the new AIM-120 missiles, the SR-71 I would be equipped with the AN/APG-65 radar used on the F/A-18 Hornet. The 27in diameter circular planar array antenna would be modified into a slightly larger 32in diameter, slightly non-circular shape. The increased area permitted detection of large aircraft out to around 100 nautical miles, with a tracking range of

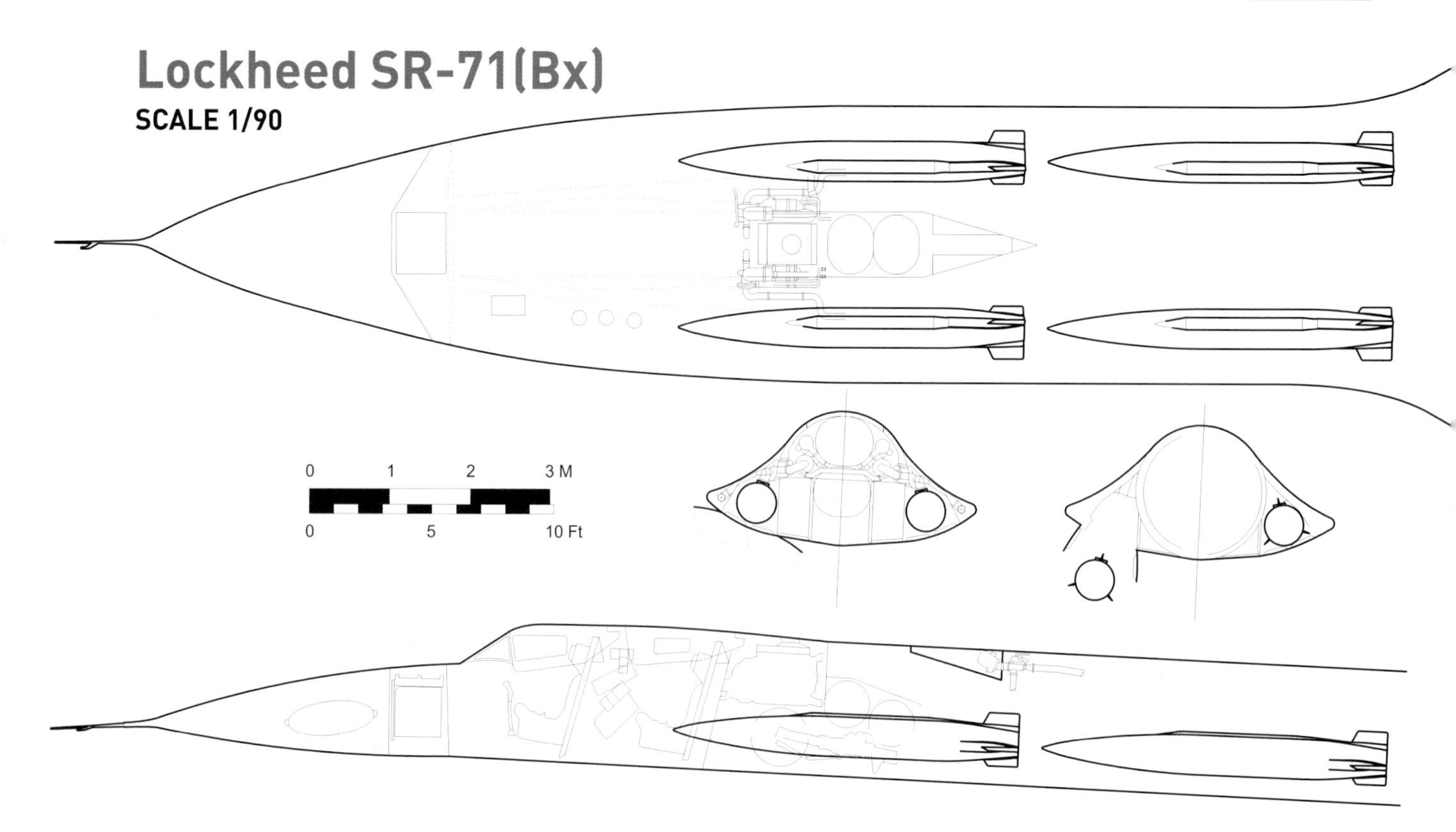

80 nautical miles. The AIM-120 missiles would have modified fins to fit within the SR-71 bays, and would have a range of about 80 nautical miles.

The purpose of the SR-71 I was to serve as a long-range high-speed interceptor of the latest generation of Soviet AWACS planes and, as a secondary purpose, to intercept Blackjack and Backfire bombers. Speed and range were such that incoming bombers could be intercepted prior to their getting close enough to the continental United States to launch their own air-to-surface missiles. It was proposed that the SR-71 I operate out of either Loring Air Force Base in northeastern Maine or Sawyer Air Force Base in Michigan (now decommissioned). From there, with a single refuelling, the SR-71 I could reach and attack Soviet targets in the region of Scandinavia and be recovered in either Thule Air Base in northern Greenland or RAF Mildenhall in Suffolk, England. One wonders whether either of these landing sites would have been anything but smoking radioactive holes by the time the SR-71 Is got there, given the likely outcome of USAF interceptors shooting down Soviet AWACS and bombers.

Lockheed did not propose to build all-new aircraft, but to modify seven existing SR-71A airframes. These seven aircraft would all be based at the same location.

HALO

For a brief moment, there was some interest in using the SR-71 to compete with the National Aero Space Plane (NASP) programme. This ambitious – as it turned out, overly ambitious – programme called for the development of a single stage to orbit aircraft powered largely by scramjet engines. The claim was that an airbreathing vehicle able to launch from and land at airport runways would prove more economical to operate as a space launcher than conventional expendable rockets. History has shown the difficulty of developing engines that work reliably and profitably with supersonic airflows to be extreme.

In lieu of the incredibly expensive and risky programme to develop the full-scale, full-speed (up to Mach 25) NASP scramjet, in 1992 NASA-Ames – otherwise not a major player in the NASP programme – proposed the development of the 'HALO' (originally the High Altitude Launch Option, later the Hypersonic Air Launch Option) research vehicle. This would be a subscale NASP-like configuration complete with scramjet engines and provisions for a crew of one. The idea seems to have lasted for a couple of years, but detailed design work has not come to light. Information on the HALO vehicle is vague and contradictory, indicating that over time it changed considerably.

The HALO vehicle was, in general, a relatively slim flat-bottomed lifting body with an underslung rectangular scramjet module. The long sloping underside of the forward fuselage served as an inlet compression ramp. At the chopped-off rear of the fuselage was a submerged linear aerospike rocket engine, needed to boost the vehicle up to a speed where the scramjet could operate effectively. The NASP vehicle is rarely depicted with such an engine, but it nevertheless would have needed rockets for the final circularization into orbit. A cockpit for a single pilot would be fitted, likely somewhat well back on the

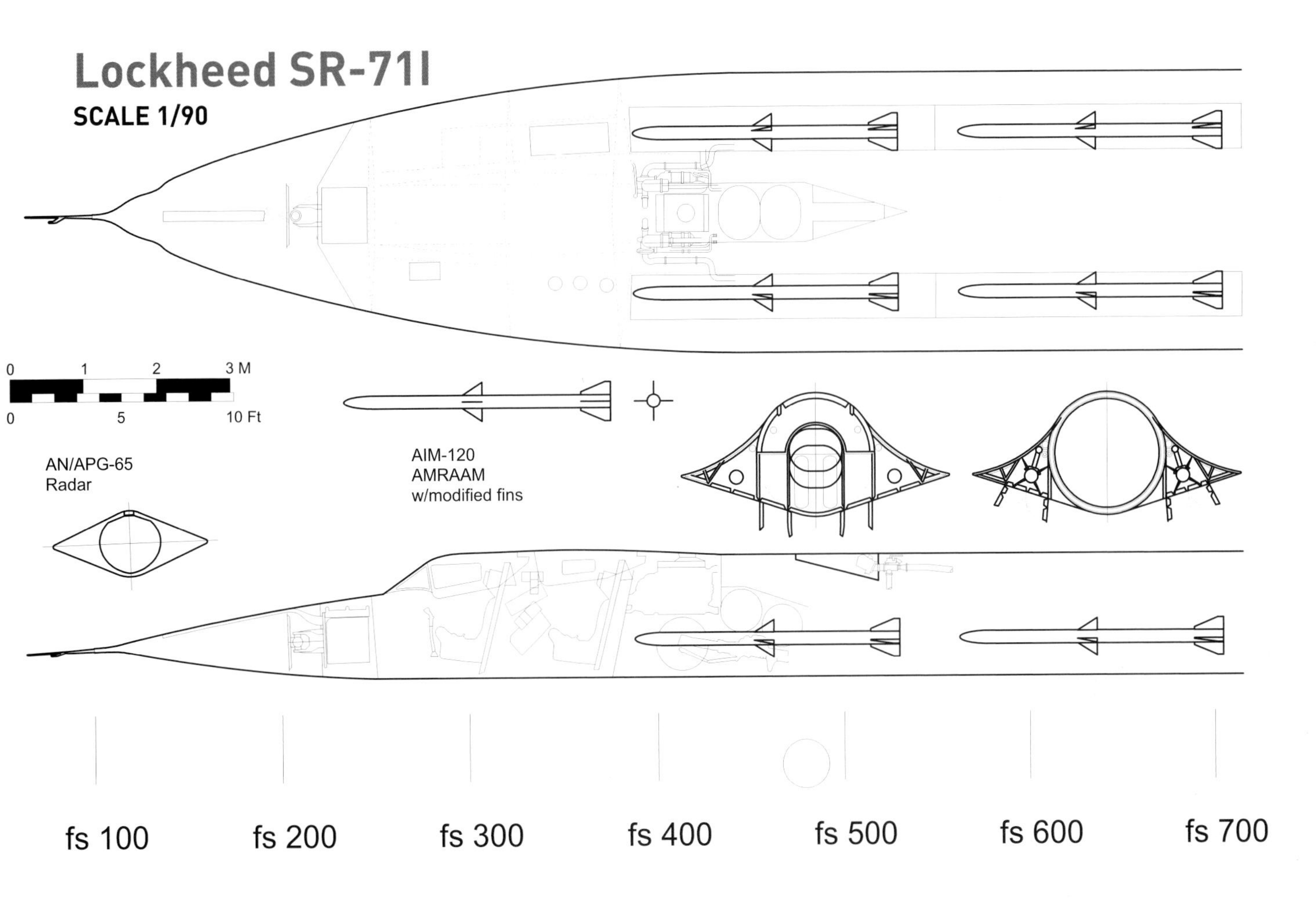

fuselage due to the relatively thin nose.

The HALO would be carried to altitude on the back of an SR-71 in a manner similar to the D-21 atop the M-21, or like the earlier Lockheed study of the HT-4 drone carried atop the SR-71. Launch of the HALO would occur at Mach 3 and 80,000ft, followed by a rocket-powered climb to Mach 9 and somewhere over 140,000ft. Following that, the craft would fly under scramjet power for two minutes, descending to around 100,000ft while accelerating to Mach 10. The pilot would then glide the vehicle to a horizontal landing in a similar manner to the X-15.

The purpose was not to aid the NASP programme, but to supplant it. By 1992, NASP was encountering serious problems with both technology and budget and the rise of a competing system only made the problems worse. NASP was finally cancelled in 1993, but HALO hung on at least into 1994.

LASRE

The last known major modification project for the SR-71 was the LASRE (Linear Aerospike SR-71 Experiment). The 1990s was a decade that saw many reusable launch vehicles proposed, with quite a number funded to a fair degree. One of these was the Lockheed VentureStar, a highly-vaunted planned single stage to orbit lifting body rocket vehicle. The VentureStar had its distant origins in lifting body launch vehicles such as the STAR Clipper of the 1960s, but more recently in the 1993 NASA Access To Space Study to which Lockheed responded with its concept of the Aero Ballistic Rocket.

A thick delta-planform lifting body, the Venture Star was equipped with a linear aerospike rocket engine system and multi-lobe composite hydrogen and oxygen tanks. The technology required to make Venture Star successful was bleeding edge and required development. Fortunately for Lockheed, NASA wanted to develop a subscale demonstrator of those sort of technologies as part of the Reusable Launch Vehicle programme. In July 1996, Lockheed's proposal for the 'X-33' was selected over designs from Rockwell and McDonnell-Douglas. Both of those companies would be consumed by Boeing before the end of the year.

The linear aerospike was the most visibly obvious of the important technologies in need of development. Such engines had been built and tested since the 1960s, but all test firings had been on the ground. None had flown. Consequently, the response of the engine and its exhaust to subsonic and especially supersonic air flows was uncertain. In order to resolve this issue NASA-Dryden Flight Research Center ran an effort to build a 20% scale model of one half of the X-33, turn it on its side and mounted it to a 'canoe' intended to be fitted to the back of a NASA SR-71.

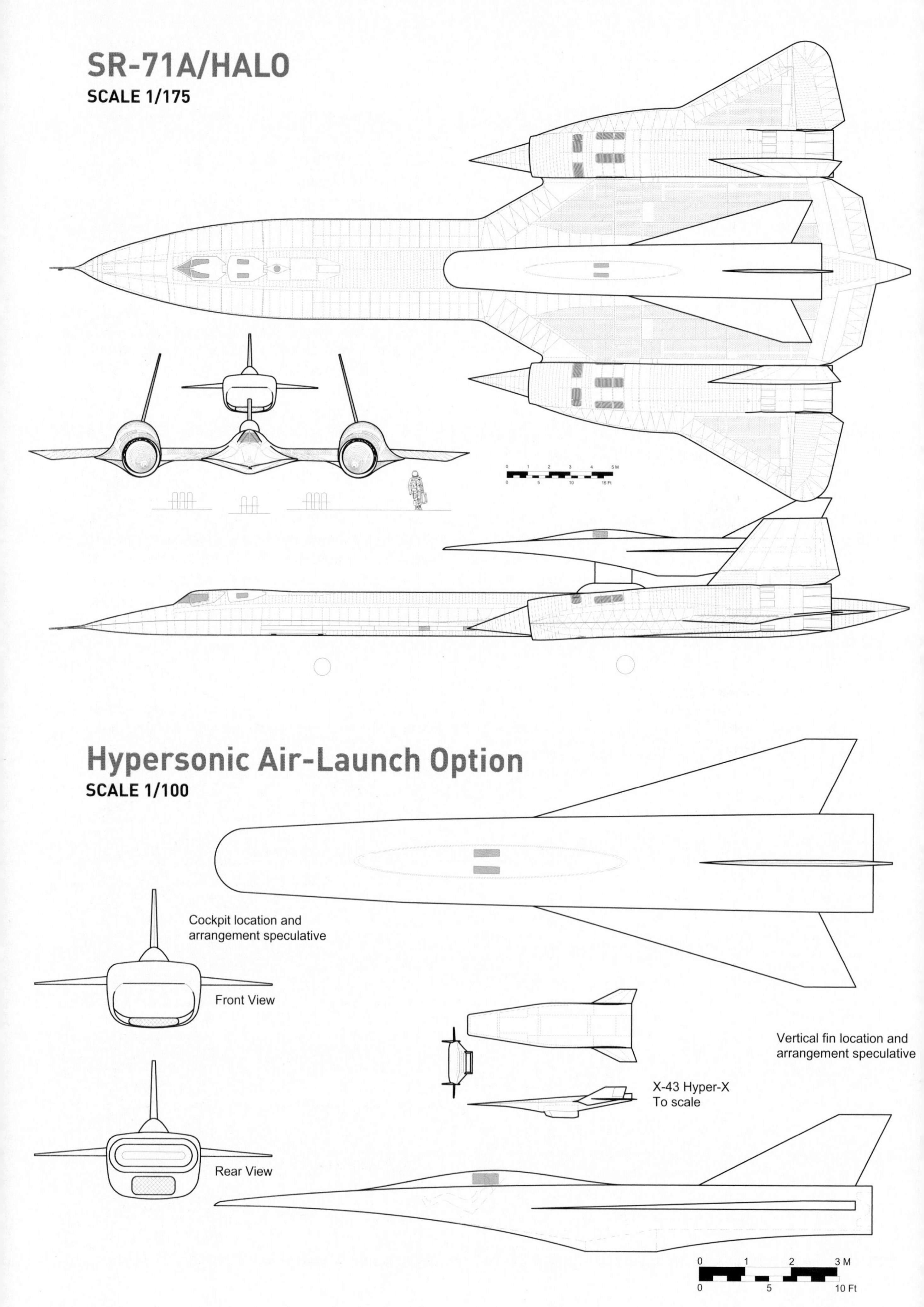
SR-71A/HALO
SCALE 1/175
Hypersonic Air-Launch Option
SCALE 1/100
Cockpit location and arrangement speculative
Front View
Rear View
Vertical fin location and arrangement speculative
X-43 Hyper-X
To scale
0 1 2 3 M
0 5 10 Ft

The LASRE would gather information during both cold-flow and hot-fire tests while flown at high Mach above an SR-71. The purpose was not to expand the speed or altitude performance of the SR-71, but simply to gather data to compare with wind tunnel tests and computational fluid dynamics simulations.

Lockheed's Skunk Works was responsible for the design, construction and integration of the overall LASRE structure (built largely of common low carbon steel, with a total gross weight of 14,500lb), while Rocketdyne built the eight-segment rocket engine thrusters which were water-cooled and made from copper. Lockheed was, unsurprisingly, also responsible for structural modifications made to the SR-71 to allow it to carry the heavy test structure on its back. The aft cockpit was modified to serve as the flight test engineer's station, providing controls for the LASRE system.

The half-model of the X-33 – not truly to scale, as the LASRE device was symmetrical (the X-33 had a bit of an airfoil shape to it) and lacked the wings and control surfaces of the X-33 – was mounted to a 'reflection plane' on top of the canoe. The reflection plane was pretty much what the name says... a flat plane, intended to fly parallel with the airstream, providing an air flow for the model much like what it would see if it were a full model flying independently. The rocket engine could produce 5500lb of thrust, burning liquid oxygen with gaseous hydrogen... but it could only do so for three seconds. Stored within the half-model itself was a tank with 335lb of liquid oxygen, while 27lb of gaseous hydrogen was stored in three tanks within the canoe. Gaseous helium pressurant moved the propellants from tanks to engine. The rocket engine would ignite with the use of TEA-TEB, something the SR-71 team was well acquainted with.

The LASRE system was built and repeatedly flown on the back of a NASA SR-71 during 1997 and 1998. However, while the rocket was ignited in static ground tests, it was never ignited in flight due to concerns over leaky oxygen lines. Liquid oxygen was cycled through the system and dumped through the thrusters; the response of the oxygen vapour to the supersonic airflow was considered informative enough to be representative of an actual hot-fire test. In any event, the X-33 programme was rife with troubles; the composite multi-lobe tank gave Lockheed no end of headaches and, along with other problems technical and financial, caused the X-33 to be cancelled outright in 2001.

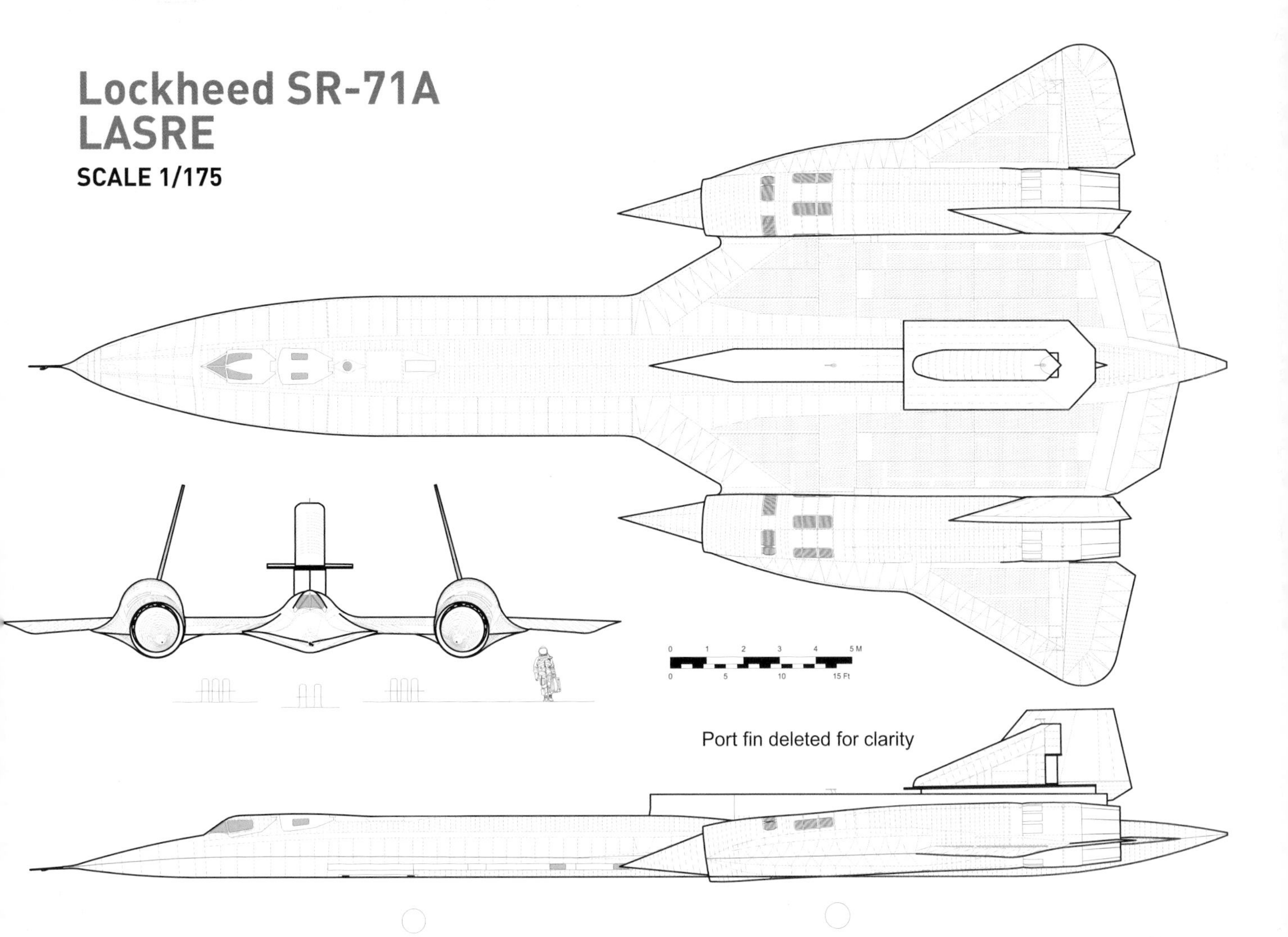

References

Chapter 1

Silverstein, A., Hall, E. 'Liquid Hydrogen as a Jet Fuel for High-Altitude Aircraft,' NACA Research Memorandum E55C28a, Lewis Flight Propulsion Laboratory, April 15, 1955
'Aircraft Configuration Survey for Weapons System 118P' North American Aviation Report No. NA-56-521, June 1, 1956
'Development Plan Report for the Special Reconnaissance Airplane Weapon System 118P' North American Aviation Report No. NA-56-464, June 4, 1956
Rich, Ben. 'Lockheed Advanced Development Projects, CL-400'
Sloop, J. 'Liquid Hydrogen as a Propulsion Fuel, 1945-1959,' NASA SP-4404, 1978
Martel, C. 'Military Jet Fuels, 1944-1987,' Aero Propulsion Laboratory, Air Force Wright Aeronautical Laboratories, Air Force System Command, AFWAL-TR-87-2062, November 1987

Chapter 2

'GEBO II Report No. 3, Structures and Weights' Consolidated Vultee report FZS-114, April 16, 1951
'Super Hustler' Convair Forth Worth presentation, report FZM-843, February 10, 1958
'Super Hustler: A new Approach to the Manned Strategic Bombing-Reconnaissance Problem,' Convair Forth Worth report FZM-1200-20, May 26, 1958
'Memorandum Report on Super Hustler Capabilities' Marquardt report MR-A-1166, May 26, 1958
'Study General Arrangement Four Ramjet Super Hustler' Convair Preliminary Design Drawing FW5810047, 7-3-58
'Status Review' Convair Forth Worth presentation, report PF-0-101M, June 9, 1959
'Kingfish Summary Report' Convair Forth Worth report, PF-0-104M, date uncertain
'Special Purpose Aircraft' General Dynamics-Fort Worth, FZM-2392, September 15, 1961
'Performance Capabilities Special Purpose Aircraft RC-182-2' General Dynamics-Fort Worth, FZM-2588, April 16, 1962
'Special Purpose Aircraft' General Dynamics-Fort Worth, FZM-2678, August 22, 1962

Chapter 3

Johnson, Clarence 'Design Study Archangel Aircraft' Lockheed SP-101, September 11, 1958
Johnson, Clarence 'A Proposal for a Lightweight Reconnaissance Aircraft' Lockheed SP-108, November 10, 1958
'Comparison Study of Proposed Follow-on Vehicle' January 1959, GUS-0086, [AES], Director of Operations
Johnson, Clarence 'Proposal A-11' Lockheed SP-114, March 18, 1959
Johnson, Clarence 'A-11 Operational Analysis' Lockheed SP-120, May 1, 1959
Burke, Colonel William, USAF 'Comparison and Evaluation of Two Weapons Systems Designed to Meet ~~CHALICE~~ OXCART Operational Requirements' June 5, 1959
Whittenbury, John R. 'From Archangel to OXCART: Design Evolution of the Lockheed A-12, First of the Blackbirds' PowerPoint presentation, August 2007

Chapter 4

'Manufacturer's Model Specification High altitude Research Airplane' Lockheed Aircraft Company, Advanced Development Projects, report SP-151, January 15, 1960
'Feasibility Study of an Air-Launched, Single-Pass, Low-Orbit Reconnaissance System' Lockheed Aircraft Company, Advanced Development Projects, report SP2-374, July 27, 1962
'Feasibility Report, Modification of A-12 Vehicle for Air-Launched Orbital Reconnaissance System' Lockheed Aircraft Company, Advanced Development Projects, report SP-404, September 7, 1962
'A-12 Manual Addendum A-12 Flight Handbook Photographic Equipment' February 1963
'Manufacturer's Model Specification (D-21)' Lockheed Aircraft Company, Advanced Development Projects, report SP-582, November 6, 1963
'D-21B Systems Description' Lockheed Aircraft Company, Advanced Development Projects, report SP-790 Vol. 1, January 1, 1968
Johnson, Clarence 'History of the OXCART Program' Lockheed Aircraft Company, Advanced Development Projects, report SP-1362, July 1, 1968
'The D-21B Reconnaissance System' Lockheed Aircraft Company, Advanced Development Projects, report SP-3066, October 10, 1968
Johnson, Kelly 'Development of the Lockheed SR-71 Blackbird' July 29, 1981
Boyer, R. and Briggs, R. 'The Use of Beta Titanium Alloys in the Aerospace Industry' The Boeing Company, The Journal of Materials Engineering an Performance, 2005
Robarge, D. 'Archangel: CIA's Supersonic A-12 Reconnaissance Aircraft' Center for the Study of Intelligence, Central Intelligence Agency, 2012
Scoville, H. Jr. 'Review of Development and Status of Project TAGBOARD' Central Intelligence Agency memo to Director of Central Intelligence, June 12, 1963

Chapter 5

'XF-103 Interceptor Technical Development Engineering Inspection' February 20, 1957, Republic Aviation Corporation
'Surfacing of RX-12' CIA memo, January 10, 1963
'Proposal for Surfacing an LRI Prototype as Cover for the OXCART Program' CIA memo, April 10, 1963
Sturdevant, C.L., Taylor, W.M. 'Utilization of the YF-12C as a Platform for Drone or Model Testing' Lockheed Aircraft Corporation, advanced Development Projects, SP-4238, September 10, 1974
Major Robert P. Lyons, Jr. 'The Search for an Advanced Fighter, a History from the XF-108 to the Advanced Tactical Fighter' Air Command and Staff College student report, Maxwell Air Force Base report no. 86-1575, April 1986
Martel, Charles 'Military Jet Fuels, 1944 – 1987' Aero Propulsion Laboratory, Wright-Patterson Air Force Base, AFWAL-TR-87-2062, November 1987

Chapter 6

Sturdevant, C.L., Taylor, W.M. 'Utilization of the YF-12C as a Platform for Drone or Model Testing' Lockheed Aircraft Corporation, Advanced Development Projects, SP-4238, September 10, 1974
Rich, Ben 'The Strategic Aspect of Supercruising Flight' Lockheed Aircraft Corporation, Advanced Development Projects, SP-4457, 2-5-1976
Johnson, Kelly 'Development of the Lockheed SR-71 Blackbird' July 29, 1981
'SR-71A Flight Manual' October 31, 1986
Corda, S. et al 'Flight Testing the Linear Aerospike SR-71 Experiment (LASRE)' NASA/TM-1998-206567, NASA-Dryden Flight Research Center, September 1998
Landis, Tony 'Big Tail – One of a Kind SR-71 Blackbird' *Wings*, December 2004

Recommended further reading:

Jenkins, Dennis *Lockheed SR-71/YF-12 Blackbirds*, Specialty Press, 1997
Suhler, Paul *From RAINBOW to GUSTO: Stealth and the Design of the Lockheed Blackbird*, American institute of Aeronautics and Astronautics, 2009
Landis, Tony *Lockheed Blackbird Family*, Specialty Press, 2010

General bibliography:

Crickmore, Paul *Lockheed SR-71: The Secret Missions Exposed*, Osprey Publishing, 1993
Miller, Jay *Lockheed's Skunk Works*, Midland Publishing, 1995
Goodall, James; Miller, Jay *Lockheed's SR-71 'Blackbird' Family*, Midland Publishing, 2002
Merlin, Peter, *Design and Development of the Blackbird*, American institute of Aeronautics and Astronautics, 2008
Davies, Steve; Crickmore, Paul *Lockheed SR-71 Blackbird: Owners' Workshop Manual*, Haynes Publishing, 2012
Crickmore, Paul *Lockheed A-12*, Osprey Publishing Ltd, 2014
Graham, Col. Richard *The Complete Book of the SR-71 Blackbird*, Quarto Publishing Group, 2015
SR-71 Flight Manual, Quarto Publishing Group, 2016
Graham, Col. Richard *SR-71: The Complete Illustrated History of the Blackbird*, Quarto Publishing Group, 2017
Goodall, Jim *Lockheed SR-71 Blackbird*, Schiffer Publishing Ltd, 2018
Dufresne, Leroy *SR-71 Handbook*, Dorrance Publishing Co. 2020

General data table

Aircraft	Image source grade	Crew	Span	Wing area (sq ft)	Length	Engines	Dry weight (lb)	Fuel (lb)	Payload (lb)	Gross start weight (lb)	Radius (n.mi.)	Cruise speed (Mach)	Cruise altitude (ft)
NACA Hydrogen Fuelled Photo Recon	1	1	~58.7ft	1,150.00	172ft	4 x 'Type C' turbojet	N/A	23,600	0	75,000	1,345	2.50	67,500
Lockheed CL-325-1	3	1	79ft 11in	2,250.00	155ft 5in	2 x Garrett REX-III	29,375	14,417	1,500	45,600	1,500	2.25	1,00,000
Lockheed CL-325-2	3	1	70ft	1,920.00	146ft 7in	2 x Garrett REX-III	34,592	14,601	1,820	41,313 (51,013)	1,210	2.25	1,00,000
Lockheed CL-400-1	4	2	83ft 9in	2,400.00	164ft 10in	2 x P&W 304-2	48,515	21,440	1,500	69,955	1,100	2.50	96,000-99,500
Lockheed CL-400-11	3	2	77ft 6in	3,000.00	206ft 8in	2 x P&W 304-2	66,508	50,000	1,500	1,16,508	2,070	2.50	88,700-95,700
Lockheed CL-400-12	3	2	110ft	6,000.00	272ft	4 x P&W 304-3	1,40,530	1,15,000	1,500	2,55,530	2,360	2.50	87,600-95,000
Lockheed CL-400-13	3	2	84ft	6,500.00	296ft 6in	2 x STR-12	2,13,150	1,62,850	1,500	3,76,000	4,494	4.00	93,000-99,400
Lockheed CL-400-14	3	2	98ft	5,500.00	290ft	4 x 85% scale STR-12	1,78,500	1,80,000	1,500	3,85,500	4,060	4.00	92,000-100,800
Lockheed CL-400-15 JP	3	2	56ft 6in	1,800.00	144ft 6in	2 x P&W J58	53,620	1,04,000	1,500	1,58,620	2,184	3.00	76,300-89,000
Bell System 118P	2	1	~67.8ft	N/A	~134ft	N/A	N/A	N/A	N/A	N/A	N/A	N/A	N/A
North American D265-27	4	1	79.9ft	6,600.00	180.9ft	4 x Aerojet ATR-2040, 103.1%	N/A	59,278	N/A	2,06,800	3,000	3.44	1,00,000
Convair GEBO II	3	2	~47.25ft	N/A	81.5ft	4 x turbojet	N/A	N/A	N/A	N/A	N/A	N/A	N/A
Convair MX-1626	3	2	47.3ft	1,200.00	72.72ft	3 x GE J53-X25	N/A	N/A	N/A	N/A	N/A	N/A	N/A
Convair Super Hustler Config. 121	4	2	23ft 4in	520.50	74ft 4in	3 x 38.6in RJ59 + 1 x GE J85	20,553	N/A	N/A	45,903	N/A	N/A	N/A
Convair Super Hustler Config. 124	4	2	23ft 6in	520.00	74ft 4in	4 x 33.5in RJ50 + 1 x GE J85	16,780	N/A	N/A	47,150	N/A	N/A	N/A
Convair 'Minimum Change' Super Hustler	3	2	19ft 10in	N/A	N/A	N/A	N/A	N/A	N/A	N/A	N/A	N/A	N/A
General Dynamics FISH Config. 220	4	1	34ft 3in	739.00	48ft 9in	2 x MA24E + 1 x P&W JT12	15,294	19,450	N/A	35,027	4,150	4.00	90,000
General Dynamics Special Purpose Super Hustler	3	1	~33ft 11in	N/A	~57ft 4in	2 x ramjet + 2 x P&W JT12	N/A	N/A	N/A	N/A	N/A	N/A	N/A
General Dynamics FISH Config. 234	4	1	37ft	714.00	47ft	2 x MA24E + 2 x GE J85	16,625	21,700	N/A	38,325	3,900	4.00	90,000
Convair FISH Config. 238	3	1	52ft 4in	N/A	69ft 5in	N/A	N/A	N/A	N/A	N/A	N/A	N/A	N/A
Convair FISH Config. 242-A	3	1	51ft 8in	N/A	78ft 11in	N/A	N/A	N/A	N/A	N/A	N/A	N/A	N/A
General Dynamics KINGFISH Config. 257	3	1	60ft	1,815.00	73ft 7in	2 x P&W JT-11	40,450	62,750	N/A	1,03,200	N/A	N/A	N/A
Convair HAZEL MC-10	4	1	67.71ft	1,985.00	N/A	N/A	6,470	6,330	N/A	30,525	3,200	3.00	1,31,400
Convair M-124A	3	1	28.42ft	500.00	56.2ft	2 x Marquardt ramjets	N/A	4,030	N/A	9,700	4,000	3.00	1,05,600
Convair P-124C	3	1	28.42ft	500.00	64.70ft	2 x Marquardt ramjets	N/A	2,920	N/A	9,100	4,000	3.00	1,04,350
Convair M-125	3	1	33.98ft	500.00	51.5ft	2 x Marquardt ramjets	N/A	5,100	N/A	11,060	4,000	3.00	1,02,300
Convair M-126	3	1	20.02ft	300.00	60.2ft	2 x Marquardt ramjets	N/A	6,900	N/A	12,190	4,000	3.00	92,200ft
Convair QRC-182-2	4	~30	58ft 11.2in	1,746.70	136ft 5.69in	4 x P&W JT-11-B2	78,570	1,12,720	10,960	2,02,250	2,700	2.40	60,000-67,000
Convair RC/80	3	N/A	57ft	N/A	125ft 6.5in	4 x P&W JT-11-B2	76,650	99,165	10,960	1,63,000	N/A	2.40	N/A
General Dynamics FISH VSF-1	3	1	55ft	N/A	47ft	2 x MA24E + 2 x GE J85	N/A	N/A	N/A	N/A	N/A	N/A	N/A
North American D265-26	4	1	64.9ft	2,757.00	120.9ft	4 x 135% GE J79-X278	83,644	1,20,399	N/A	2,07,800	3,032	2.75	75,000
North American Manned SM-64	1	1	42.75ft	761.00	87.3ft	N/A	N/A	2,41,070	N/A	3,00,152	3,950	N/A	53,350-71,800
Lockheed Archangel	3	1	~47.5ft	N/A	~119.1ft	2 x P&W J58	41,000	43,000	N/A	N/A	2,000	N/A	86,000
Lockheed A-1	5	1	~59.9ft	1,650.15	116ft 8in	2 x P&W J58	41,000	61,000	1,090	1,02,000	4,022	3.00	83,000-93,000
Lockhed Peterbilt	3	1	110ft	N/A	N/A	4 x P&W J75	N/A	N/A	N/A	N/A	N/A	0.80	70,000
Lockheed Kite	4	1	134ft 8in	6,700.00	102ft 2in	1 x 180in ramjet	N/A	N/A	800	20,000	4,000	N/A	N/A

Aircraft	Image source grade	Crew	Span	Wing area (sq ft)	Length	Engines	Dry weight (lb)	Fuel (lb)	Payload (lb)	Gross start weight (lb)	Radius (n.mi.)	Cruise speed (Mach)	Cruise altitude (ft)
Lockheed A-2	5	1	76.76ft	2,486.50	129.17ft	2 x P&W J58 + 2 x 75in ramjets	53,125	81,000	500	1,35,000	4,000	3.20	100,000-105,000
Lockheed Cherub 1 A-3	4	1	38.25ft	497.00	58.05ft	N/A	N/A	N/A	N/A	N/A	N/A	N/A	N/A
Lockheed Cherub 2 A-3	4	1	28ft 6in	509.00	58ft 9in	N/A	N/A	16,000	N/A	N/A	N/A	N/A	N/A
Lockheed Cherub Variant A-3	4	1	38ft 3in	497.00	64ft	N/A	N/A	N/A	N/A	N/A	N/A	N/A	N/A
Lockheed A-3 Variant	4	1	37ft 10in	606.00	52ft 2in	N/A	13,930	14,000	0	27,930	N/A	N/A	N/A
Lockheed A-3	4	1	33ft 4in	500.00	68ft 10in	N/A	N/A	20,860	N/A	32,800	N/A	N/A	N/A
Lockheed A-4-2	4	1	35ft w/o ramjets	N/A	58.33ft	1 x P&W J58 + 2 x 34in ramjet	N/A	N/A	N/A	N/A	1,320	N/A	92,000
Lockheed A-5	4	1	32ft 6in	750.00	46ft	1 x H2O2 rocket engine + 1 x 83in ramjet	N/A	N/A	N/A	N/A	1,557	3.20	90,000
Lockheed A-6-5	4	1	47ft 2in	1,275.00	64ft	1 x P&W J58 + 2 x 32in ramjet	N/A	N/A	N/A	N/A	1,287	3.20	90,000
Lockheed A-6-6	4	1	47ft 2in	1,075.00	62ft 10in	1 x P&W J58 + 2 x 34in ramjet	N/A	N/A	N/A	N/A	N/A	N/A	N/A
Lockheed A-6-9	3	1	47ft 4in	1,118.00	68ft 10in	1 x P&W J58 + 2 x 34in ramjet	N/A	50,000	N/A	N/A	N/A	N/A	N/A
Lockheed Arrow	3	1	25ft 6in	700.00	55ft	1 x P&W JT-12 + 2 x 40in ramjet	N/A	N/A	N/A	N/A	N/A	4.00	90,000
Lockheed 'B-58 Launched Vehicle'	4	1	31ft 7in	700.00	44ft 7in	1 x P&W JT-12 + 2 x 40in ramjet	N/A	N/A	N/A	N/A	N/A	4.00	95,000
Lockheed A-7-2	4	1	42.125ft	766.35	78.33ft	1 x P&W J58 + 2 x 34in ramjet	N/A	N/A	N/A	N/A	N/A	N/A	N/A
Lockheed A-7-3	4	1	47ft 6in	989.58	93ft 9in	1 x P&W J58 + 2 x 34in ramjet	N/A	N/A	N/A	N/A	N/A	N/A	N/A
Lockheed A-10	5	1	46ft	1,400.00	109ft 6in	2 x GE J93	32,315	N/A	N/A	86,000	2,000	3.20	90,500
Lockheed A-11A	5	1	53ft	1,404.50	105ft 5in	2 x GE J93	N/A	N/A	N/A	N/A	N/A	N/A	N/A
Lockheed A-12 Initial Design	4	1	56ft	N/A	100ft	2 x P&W J58	N/A	N/A	N/A	N/A	3,940	3.20	83,000
Lockheed A-12 Canard	2	1	N/A	N/A	N/A	N/A	N/A	N/A	N/A	N/A	N/A	N/A	N/A
Lockheed A-12	3	1	55ft 7in	1,795.00	102ft 3in	2 x P&W J58	N/A	N/A	N/A	1,17,000	3,400	3.20	90,000
Lockheed AP-12	3	2	55.617ft	1,795.00	98.75ft	2 x P&W J58	N/A	N/A	N/A	1,17,000	N/A	3.20	N/A
Lockheed D-21B	5	0	19.08ft	388.50	42.85ft	1 x Marquardt RJ-73	5,130	5,800	N/A	10,330	3,000	3.30	80,000-95,000
Tupolev Voron	4	0	19.02ft	398.27	42.85ft	1 x RD-012 ramjet	3,450	2,850	N/A	13,889	2,484	3.25-3.55	75,459-86,614
Lockheed Little Harvey Concept A	3	1	13ft	N/A	26ft	N/A	N/A	N/A	N/A	5,000	N/A	N/A	N/A
Republic XF-103 (Jan '57)	3	1	44.65ft	401.00	81.9ft	1 x Wright YJ67-W-3 + 1 Wright XRJ55-W-1	32,757	N/A	N/A	55,780	N/A	2.24-3.00	75,000
North American F-108	4	2	57.58ft	1,065.00	88.75ft	2 x GE J93	50,907	47,632	N/A	1,02,533	1,010	2.58	N/A
Convair B/J-58 LRI	3	2	56ft 10.3in	1,780.00	108ft 9.5in	2 x P&W J58	N/A	N/A	N/A	1,39,000	N/A	2.40	75,000
Lockheed AF-12	3	2	59ft	1,795.00	101.66ft	2 x P&W J58	N/A	N/A	N/A	1,24,000	N/A	N/A	N/A
Lockheed YF-12A	3	2	55ft 7in	1,795.00	101.8ft	2 x P&W J58	N/A	N/A	N/A	1,24,000	3,000	3.20	N/A
Lockheed AF-112D	3	2	59ft	1,795.00	103.89ft	2 x P&W J58	N/A	N/A	N/A	N/A	N/A	N/A	N/A
Hypersonic Air-Launch Option	3	1	17ft	N/A	47ft	N/A	N/A	N/A	N/A	N/A	N/A	N/A	N/A

Every drawing in this volume is based on original primary source material. The key to image source grading is as follows:
1 The drawing is provisional, based on a text description.
2 The source drawing is at best crude, or an isometric or perspective artist's impression or photos of a model.
3 The source drawings are serviceable but simple.
4 The source drawings were clear, but the design was not entirely detailed.
5 The source material was detailed, clear and unimpeachable.